APOPTOSIS

The concept of programmed cell death, or apoptosis, has exploded into a major scientific field of interest for cell biologists, oncologists, and many other biomedical researchers. Apoptosis occurs throughout the lifetime of most multicellular organisms. During development, for example, the selective death of cells is vital to remove tissue between the digits to produce fingers and toes. Apoptosis is also necessary to destroy cells that represent a threat to the integrity of the organism, for example, cells infected by a virus. In many cancers the genes regulating apoptosis are defective, producing immortal, continuously proliferating cells. This book discusses the philosophical and technical difficulties in defining the moment of death for a cell, as well as the biological implications and significance of programmed cell death. Recent developments in the genetic control and interacting gene networks associated with apoptosis are presented. The book is written for advanced undergraduate and postgraduate students, and is highly illustrated to aid understanding.

Christopher Potten is Chairman and Scientific Director of EpiStem Ltd. and an Honorary Professor of Stem Cell Biology at the University of Manchester. He worked for many years in cancer research.

James Wilson is a Lecturer in Paediatric Gastroenterology at Bart's and The London, Queen Mary's School of Medicine and Dentistry.

Apoptosis

THE LIFE AND DEATH OF CELLS

Christopher Potten
EpiStem Ltd., Manchester

James Wilson
Queen Mary, University of London

CAMBRIDGE UNIVERSITY PRESS
Cambridge, New York, Melbourne, Madrid, Cape Town, Singapore, São Paulo

Cambridge University Press
40 West 20th Street, New York, NY 10011-4211, USA

www.cambridge.org
Information on this title: www.cambridge.org/9780521622714

© Christopher Potten and James Wilson 2004

This book is in copyright. Subject to statutory exception
and to the provisions of relevant collective licensing agreements,
no reproduction of any part may take place without
the written permission of Cambridge University Press.

First published 2004
Reprinted 2005

Printed in the United States of America

A catalog record for this publication is available from the British Library.

ISBN-13 978-0-521-62271-4 hardback
ISBN-10 0-521-62271-9 hardback

ISBN-13 978-0-521-62679-8 paperback
ISBN-10 0-521-62679-X paperback

Cambridge University Press has no responsibility for
the persistence or accuracy of URLs for external or
third-party Internet Web sites referred to in this book
and does not guarantee that any content on such
Web sites is, or will remain, accurate or appropriate.

In memory of Sarah

Contents

Preface	*page* xi
Acknowledgements and dedications	xv

1 Dead or alive — 1
- Movement — 1
- Metabolism — 2
- Sensory perception — 3
- Reproduction — 3
- Cell death: human analogies — 7

2 How to die — 15
- The undead — 15
- The clearly dead — 17
- Necrosis — 18
- Apoptosis — 20
- Situations where death might be initiated — 23
- How long to die? — 28
- Occurrence of apoptosis — 29
- What's in a name? A rose is a rose... — 33
- Mitotic death — 36
- Apoptosis versus necrosis — 37
- DNA degradation — 38
- How do we recognise apoptosis? — 42
- Assessment of DNA fragmentation — 43
- Assessment of protease activity in apoptosis — 49
- Changes in apoptosis-regulatory proteins — 52

Membrane changes	54
Morphology	56
Cell death in cell cultures	57

3 What to wear and who clears up the rubbish? — 61

4 To reproduce or die? — 67

Defining our terms	67
DNA replication	69
Cell division	71
How do we recognise a proliferating cell?	75
Recognition of cells replicating their DNA	77
Cell cycle quiescence	84
Cyclins	86
Flow cytometry techniques	88

5 The judge, the jury, and the executioner – the genes that control cell death — 91

p53 – The guardian of the genome in embryos and adults	91
Genes that determine survival or death – the *bcl-2* family	98
Apoptotic proteases	103
The big picture	107

6 Stem cells — 115

What is a stem cell?	115
Stem cell definition	120
Stem cells and tissue injury	124
Self-maintenance probability	126
A test of functional competence for stem cells: clonogenic cells	126
Are stem cells intrinsically different from transit cells?	129
Differentiation options: pluripotency	130

7 An *in vivo* system to study apoptosis: the small intestine — 136

Proliferative organisation in the gut	136
Apoptosis in the gut	151

CONTENTS

Apoptosis induced by other cytotoxics: are all cells
programmed to die? ... 162
Apoptosis in the large bowel and the role of *bcl-2* ... 164
The *adenomatous polyposis coli* (*APC*) gene ... 167
Small and large intestine cancer incidence figures ... 169
Genome protection mechanisms ... 170

8 Cell death (apoptosis) in diverse systems ... 184

9 Measuring the levels of cell death (apoptosis) ... 189

Index ... 199

Preface

The death of cells in tissues and organisms was recognised by nineteenth-century histologists and anatomists, particularly those from Germany. On the whole it was regarded as a relatively unimportant process; an entirely passive phenomenon that occurred as a consequence of individual cells sustaining damage or becoming senescent and dying. This view remained unchanged for more than a century until it was realised that the death of cells in a tissue was part of the counterbalance to cell division in determining the overall rate of growth of the tissue. This came to prominence in studies on the growth of tumours, particularly from the work of Gordon Steel, where cell loss was realised, in addition to cell proliferation, to be important in contributing to the overall rate of tumour growth. There are two major elements that counterbalance proliferation. First, there is loss of proliferating cells to a functionally differentiated state with no capacity to return to a proliferative state. Second, there is loss of cells through cell death. Dead cells were recognised in sections of tumours analysed through the microscope in a way that was similar to that described in the early literature. For a time, the concept of cell death as an important factor in growth rate remained the almost exclusive preserve of those working on questions relating to tumour growth. From the early days of microscopic analysis, researchers of the embryological development of organisms from worms and flies to humans, and those studying metamorphosis in the life cycle of organisms such as butterflies and frogs, realised that the loss or the removal of cells was a vitally important phenomenon: this process of cell removal was termed *programmed cell death*.

PREFACE

In 1972, a group of pathologists from Aberdeen and Brisbane published a research paper that radically changed the field of research into the processes of cell death. In this paper, they described in detail the changes that occurred in the electron microscopic appearance of individual cells when they died: in this case, the death of kidney tubule cells in response to high levels of a corticosteroid. This and subsequent studies clearly indicated that, far from being a passive phenomenon, cell death could involve considerable cellular activity. The term *apoptosis* was coined to describe this process of cell death, with active cellular involvement implying a suicide-like process was programmed into cells. There was initial reluctance by the more conservative elements of the scientific community to accept the phenomenon of apoptosis, and for a while programmed cell death, as described by developmental biologists, and apoptosis were thought of as separate processes. Inevitably though, these have now been seen to be essentially the same active process, and since the mid-1980s apoptosis research has flourished into an exponentially expanding field. Surprisingly, it was work in the United Kingdom and Australia that kept the concept of apoptosis alive in the late 1970s and early 1980s. It was only after this that the American scientists entered the field.

It is now recognised that cells may die because they become old and defective, because they are surplus to the requirements of the tissue, or because they incur some damage. Each of these possibilities involves considerable internal cellular programming to regulate gene and protein expression. In addition to internally derived signals, cells can be instructed to commit suicide in response to external signals from their direct neighbours or local cells, from cells of the immune system, and from systemically derived signals.

Apoptosis is an integral part of the regulation of tissue morphogenesis during development and also the regulation of cell production under the stable conditions that one sees in adult organisms in species as diverse as worms, flies, mice, and humans. Abnormalities of tissue growth (e.g., shrinkage or atrophy of tissues with ageing, and diseases of increased proliferation like psoriasis and abnormal

PREFACE

growth like cancer) may all result from an imbalance between the processes of cell division and cell loss by differentiation and apoptosis. Extensive studies are currently underway to determine the molecular and genetic regulation of apoptosis in a variety of situations. Such studies may help in understanding and ultimately preventing the development of tumours and other diseases of proliferation. They may facilitate also the development of entirely new strategies for treating such diseases.

With the evolution and expansion of research into apoptosis, it is becoming an important discipline within the field of cell biology. I have attempted here to describe in simple terms the current status of our knowledge of apoptosis, to explain some background to the field, and to describe some of the difficulties and uncertainties that surround experiments involving apoptosis and the attempts to assess quantitatively the number of cells dying in various circumstances. I have not included extensive lists of references, which clutter and distract from reading the text, but have included various lists of additional reading matter, which generally are either key papers or extensive review articles from which further papers may be identified.

Most of the conclusions that one draws in scientific research represent approximations to the truth. This is particularly true of the field of apoptosis research as it stands at the moment. The search for the truth is, in my view, not particularly facilitated by some of the modern techniques of molecular biology, which commonly are performed on highly specialised cell culture systems and often make use of unconvincing changes in the intensity of blots and gels and make little use of statistical approaches for testing for significance. In apoptosis research, the work can often be based on single or a small number of experiments, the results of which are rarely, if ever, of the 'all or none' type.

It is relatively easy to identify cells that are dying in tissues, but to define the moment at which cell death starts and finishes is extremely difficult. The problems of defining life and death for a cell are similar to the current medical and ethical difficulties associated with defining life and death in humans. Furthermore, the number of cells that one sees displaying the characteristics of dead cells in a tissue may not be the same thing as the number of cells dying per unit of time, for a variety of complex reasons.

I have been involved in apoptosis research since the mid-1970s and my interests have centred on the role played by apoptosis in a rapidly dividing tissue: the intestinal mucosa. This is probably one of the most extensively studied tissues and provides a nice model biological system for studying cellular interactions. I shall refer to it fairly heavily and because it is a tissue with which I am very familiar it will, inevitably, be something of a personal and idiosyncratic view that is presented.

I have also attempted to explore some of the difficulties outlined above, which on the whole have not been addressed by current researchers. I do this to emphasise the point that nothing is entirely clear and conclusively resolved in science, in the hope that it may stimulate new, young scientists to attempt to address some of the questions and clarify the uncertainties, and so approach closer to the scientific truth. I also hope that some may become interested in understanding the complex biochemical interactions and intercellular dialogues that go on between cells in the body that determine whether they divide, differentiate, or die.

C.S.P.

Note: A good background review on apoptosis can be found in Harmon, B. V., and Allan, D. J., Apoptosis: a 20th century scientific revolution. In: *Apoptosis in Normal Development and Cancer*. Taylor & Francis. London, 1–19, 1996.

Acknowledgements and dedications

I have been thinking of setting down a personal view of the topic of apoptosis for some time and I am grateful to Cambridge University Press for giving me the opportunity to do so.

I would like to thank my friend and colleague Professor Aldo Beccioloini for providing me with a desk, some peace and quiet, a beautiful climate, and a cultural, gastronomic, and oenologistic environment in Florence, in which I was able to put together many of these chapters whilst a visiting professor over several years. Dr. Manuela Balzi has also always been an invaluable friend and assistant while I have been in Florence.

I should also like to thank Professor John Kerr and his colleagues, particularly Jeffrey Searle, Brian Harmon, and David Allan, for introducing me to the topic of apoptosis in the early seventies and inviting me to stay in Brisbaine as a visiting professor some years ago.

Throughout my scientific career at the Paterson Institute for Cancer Research, I have been supported by long-term grants and a life fellowship from the Cancer Research Campaign (now Cancer Research UK), to whom I am extremely grateful.

I am grateful to Christine Sutcliffe for her help over the past few years in preparing many documents and reports for a largely computer-illiterate scientist.

A book like this would not be possible without many years of work in the field, during which I have relied on the hard work of a large number of dedicated and loyal staff and visiting scientists performing the necessary experimental work. These are too many to list but I make an exception for my long-suffering research assistants,

Caroline Chadwick, who worked with me for more than twenty years, and Dawn Booth, for her highly professional approach and help. I also want to acknowledge Dr. Cath Booth for her friendship, help, and continuous good humour during the various highs and lows of academic grant-funded research. It is to Caroline, Cath, and Dawn and to all the friends and colleagues who have supported me over the years that I dedicate this book. I would also like to acknowledge the love, great help, and support of Ed, Steve, and Mike over the past few difficult years.

C.S.P.

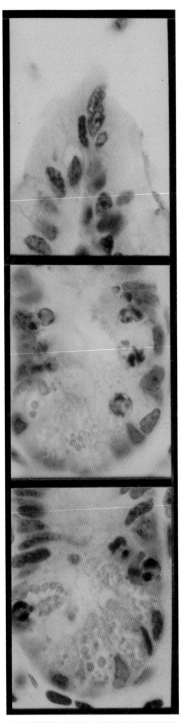

I. TUNEL assay for apoptosis in intestinal epithelium. Upper panel shows the functional cells at the tip of the villus (see Chapter 7). The lower two panels show apoptosis in the proliferative compartment, the crypts. Cells may be observed with characteristic apoptotic morphology (arrowheads). Most of these cells also show brown stain (TUNEL). Cells in division (mitosis) are shown by arrows.

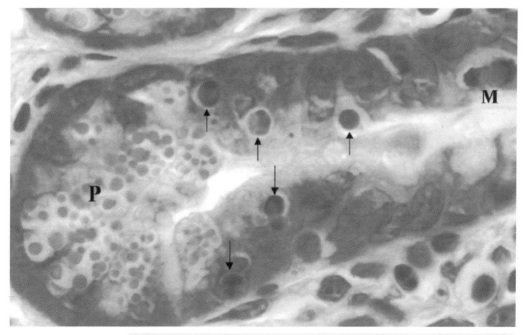

II. Haematoxylin-and-eosin-stained section through a small intestinal crypt showing many apoptotic fragments (arrows), a mitotic cell (M), and functional differentiated cells, Paneth cells (P).

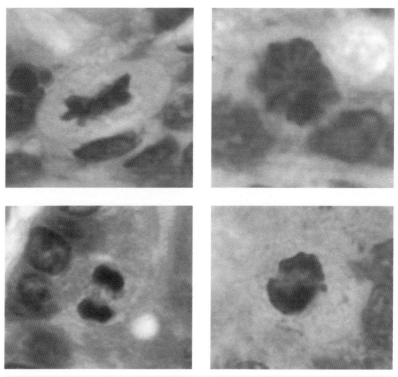

III. Mitotic figures in the small intestinal crypt (haematoxylin and eosin). Top two figures show metaphase plates; bottom two figures show anaphases.

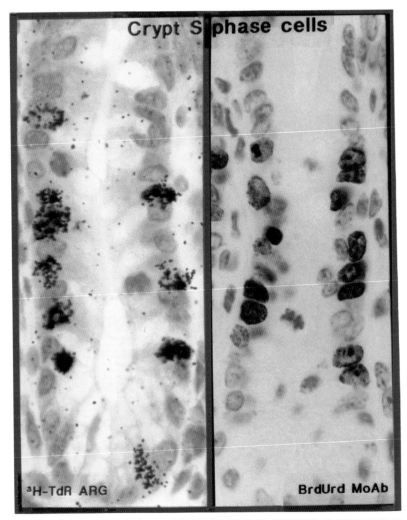

IV. Sections through small intestinal crypts showing cells replicating their DNA, either by tritiated thymidine labelling and autoradiography or by bromodeoxyuridine labelling and immunohistochemistry (mitotic cells indicated by arrows).

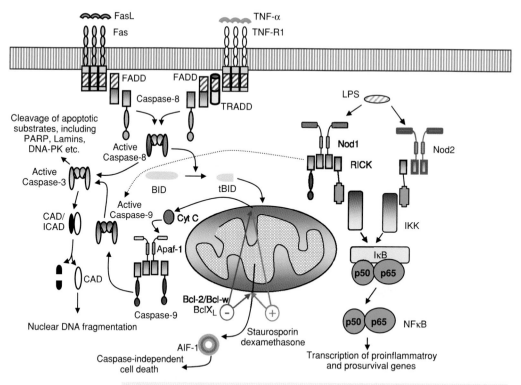

V. Diagram showing both signalling pathways for induction of cell death via death receptors and via stimuli that result in mitochondrial membrane permeability transition. The interaction of the two pathways should be noted.

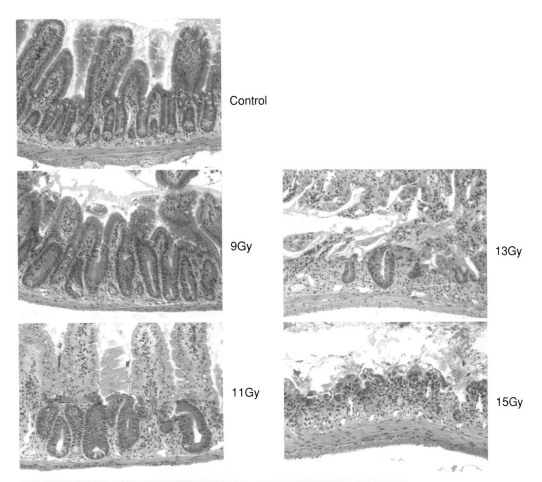

VI. Haematoxylin-and-eosin-stained sections of mouse small intestine (control and four days after various doses of radiation). In the control, the crypts and the villi can be clearly seen. In the postirradiated samples, the regenerating crypts can be seen. They decrease in number as the dose is increased. They are larger in size at day four than the control crypts.

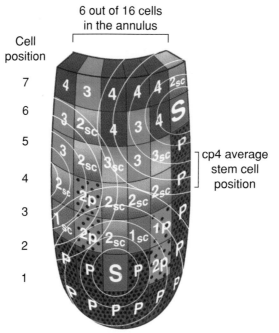

P = Paneth cell
1p, 2p = Paneth lineage cells
S = Stem cell
1_{sc} = 1st transit lineage clonogenic stem cell
2_{sc} = 2nd transit lineage clonogenic stem cell

VII. Diagram of the small intestinal crypt showing the spatial distribution of the lineage ancestor stem cells. The numbers of these crucial cells per crypt must be tightly regulated or the crypts would vary enormously in size, because each stem cell generates a lineage of up to 128 daughter cells. The regulation in numbers may be determined by the levels of a stem cell signal, such as illustrated by the concentric rings on the diagram. Reproduced with permission from Marshman *et al.*, 2002.

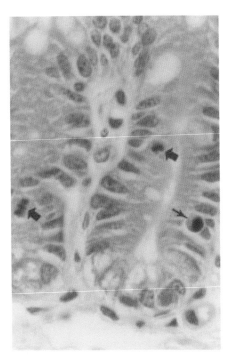

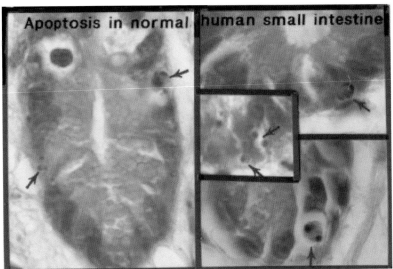

VIII. (Top panel) Haematoxylin-and-eosin-stained section of a normal, healthy mouse showing cells in mitosis (large arrows) and a single apoptotic cell (small arrow). (Bottom panel) Similar spontaneous apoptosis can also be observed in normal, healthy human small intestine.

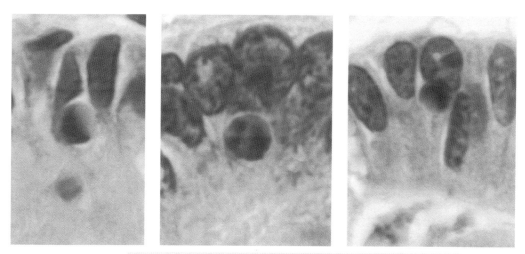

IX. Three examples of apoptosis near the base of the small intestinal crypt (haematoxylin and eosin staining).

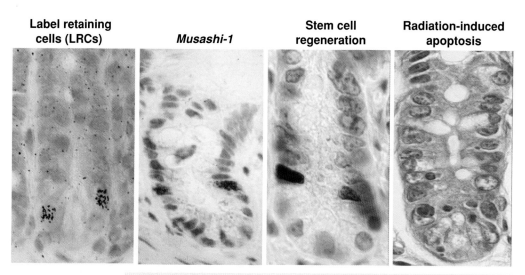

X. Four approaches to studying small intestinal stem cells. Labelling the template strands according to the Cairns's hypothesis generates label-retaining stem cells; antibody staining to *musashi-1* identifies early lineage cells (including stem cells) and the radiation-induced apoptosis is a response specifically related to the stem cells. The early stages of postinjury regeneration are believed to involve stem cells. Twenty-four hours after a high dose of a chemotherapeutic agent, a few cells at the stem cell position in the crypt enter DNA synthesis and can be labelled with bromodeoxyuridine.

1

Dead or alive?

How do we know whether a cell is dead or alive? This question is somewhat analogous to the difficult ethical and philosophical questions that doctors face in defining the point of death of a human. In order to consider this question we need first to determine what we understand by the term *living*. Conventionally, life is generally defined by an organism being able to demonstrate various activities. Broadly speaking, these can be divided into four categories as follows:

(a) movement,
(b) metabolism,
(c) sensory perception, and
(d) reproduction.

These properties shall now be considered individually.

Movement

Most living organisms exhibit movement relative to their environment, to a greater or lesser degree. There are some apparently sedentary organisms, but even in these some levels of movement are generally detectable. For example, many plants, although fixed at a particular position in space display phototropism and can track the movement of the sun throughout the day. Plants are also geotropic and their roots grow (move) downwards in response to gravity, whereas their shoots show phototropism and grow upwards in response to light. Within the mammalian body, considerable evidence of individual cellular movement can be found. This ranges from the movement of cells through tissues as they mature and differentiate into functional cells to the passive movement of circulating red blood cells to active migration of leukocytes into and within tissues in response to chemotactic signals to, of course, the highly active and directional swimming of sperm in response to chemical signals.

In addition to this movement of organisms in three-dimensional space and the movement of cells within the three-dimensional structures of tissues, there is clear evidence of intracellular movement, that is, the movement of organelles within the body of the cell (seen clearly in plants, fungi, and animals). These movements, such as the 'flow' of mitochondria through the cytoplasm, can be readily observed using time-lapse video microscopy. This type of movement within a cell is somewhat analogous to the movement of blood through the blood vessels in a multicellular, multitissued animal. Movement, whether of individual cells or whole organisms, clearly involves extensive coordination of inter- and intracellular mechanisms and cellular skeletal elements that permit movement. These require biochemical reactions and ultimately the transcription and translation of genes.

Metabolism

All living organisms, including individual cells in the tissues of the body, have to undertake biochemical, metabolic processes in order to convert nutrient materials into useful cellular products and to process waste by-products for excretion. In humans, this metabolic activity requires oxygen uptake and utilisation, with carbon dioxide being produced as a waste product, whereas plants take up and utilise carbon dioxide and excrete oxygen. This is one clear reason why we are so completely dependent on our botanical co-inhabitors of this planet. Cellular metabolism provides essential materials (molecules) required for the correct functioning of the cell, tissue, or organism as a whole and for an individual cell to grow and increase its mass or for cells to divide and multiply to increase the mass of a tissue, that is, for tissue or body *growth*. It is relatively easy to see how these metabolic activities can be considered part of a definition of life for individual cells, as well as multicellular organisms like humans.

Sensory perception

Another fundamental property of life is the ability to perceive and respond to environmental signals. Plants respond to light, gravity, temperature, nutrient and water supplies. Animal cells and organisms respond to much the same types of stimuli. Therefore, both cells and organisms must have mechanisms for detecting the environment (i.e., sensory cells or tissues) and an ability to respond to the changes that they detect. Once again, these processes require the coordination of complex intra- and extracellular programmes.

Reproduction

The ability of cells to reproduce themselves is fundamental to the life of multicellular organisms and the ability of whole organisms to reproduce (divide) is essential to the survival of all life-forms. Bacteria may reproduce by simply dividing in two. In more complex organisms such as mammals, cells reproduce in a similar way, by a process called *mitosis*. Increasing cell numbers through division contributes to the growth of tissues, along with increased mass and volume of individual cells.

Most living organisms also reproduce sexually, by producing a special form of reproductive or germ cell by a special type of cell division, termed *meiosis*, which results in the production of male and female germ cells, sperm and ova, each of which has half the normal, full cellular DNA content (i.e., for humans a single copy of each of the twenty-three chromosomes). When germ cells are brought together, they fuse to form a new, single-celled organism (the fertilised zygote) from which the highly complex multicellular organism will develop via repeated acts of cell division. Before mitotic cell division can occur the genetic material of the cell, the DNA, must be replicated so that immediately prior to mitosis, the cell contains a double complement of DNA (four copies of each chromosome for humans).

In addition to sexual reproduction, there are other forms of vegetative or nonsexual procreation. These may involve the process of budding off a new individual or other acts that amount to *cloning*, where offspring are generated that are identical to the parent (i.e., the growth of new plants or animals from severed segments (cuttings) of the parent. Such asexual reproduction fails to generate genetic diversity and as a consequence is not beneficial in adverse or altered environmental conditions where natural selection is advantageous.

As stated previously, for a cell to reproduce itself (i.e., divide) it must first reproduce its genetic material, the DNA, and other essential, subcellular elements. Each human cell has about two metres of DNA coiled within its nucleus. This amount of DNA can be accommodated within a cell's nucleus by virtue of its specialised structure. DNA has a double-helix structure and is complexed and coiled up with other molecules (these include proteins called *histones*) to form chromosomes. In humans there are twenty-three different chromosomes, with most cells having one pair of each (i.e., forty-six chromosomes). The genetic code relies on the appropriate arrangement of the DNA bases, adenine, guanidine, cytosine, and thymidine (A, G, C, and T, respectively), which form the coded messages of the individual genes that control all cellular activity. To copy the genetic code of the DNA the chromosomes must be uncoiled (an extremely complex and not fully understood process). The synthesis, or replication, of DNA is very complex and must be completed without errors, which would otherwise be multiplied in the successive generations of cells: genetic errors are the basis of diseases such as cancer. The DNA must then be recoiled and condensed back into the chromosome structures that may then be separated by the process of cell division or mitosis. These sequential processes of DNA uncoiling, replication, recondensation, and cell division are collectively termed the *cell cycle* and are described in greater detail in Chapter 4 of this book.

The ability of a cell to reproduce itself via division (mitosis) is commonly used as a measure of cell viability in many studies where toxic drugs, chemicals, or radiation are, for example, assessed for their ability to kill cancer cells. This cellular reproduction is alternatively referred to as cellular proliferation. A cell may fail to divide successfully because the DNA, or other critical elements of the cell, have been damaged in such a way as to prevent DNA replication or cell division. The damaged cell may then undergo degradative changes. If the damage is particularly severe the cell will swell and lyse, spilling out its contents, a process of cell death called *necrosis*. Alternatively, the cell may activate a programme of genetic and biochemical changes to control its own death – suicide, if you like; this process is termed *apoptosis*. This involves shrinkage of the cell and its fragmentation. The fragments of the cell may then be engulfed and digested by other cells. Genetically programmed and temporally controlled cell death, termed *programmed cell death*, was well recognised in developing embryos prior to the description of apoptosis in the scientific literature. Apoptosis and programmed cell death are essentially variations on a theme and the terms are used interchangeably throughout many scientific publications.

It may be that a cell will attempt to undergo mitosis but because of damage to the DNA (or other cellular structures) it is incapable of exiting this phase of the cell cycle and by default dies by apoptosis. This is a process we shall call here *mitotic cell death* (see 'Mitotic death' in Chapter 2).

Finally, cells may also fail to divide when they switch from proliferation to *differentiation*, that is, they mature and take on a specific functional role. The fact that a fully differentiated, functional cell does not replicate its DNA and does not divide means that a certain level of DNA damage may be tolerated, as the stringent checks for DNA errors that occur during the cell cycle are not applied or needed. Differentiation can be regarded as the process, or series of steps, that leads to the production of a functional or differentiated cell (defined in greater detail in Chapter 4). Typical differentiated cells are red blood cells, the cells on the outer surface of your skin, muscle and nerve cells, and cells that secrete products such as mucus, enzymes, saliva, and milk.

Full or terminal differentiation is usually incompatible with further rounds of cell division. Cells that do not obviously fall into this category may, nevertheless, express a spectrum of limitations in terms of their division potential. These may range from their ability to divide once after exposure to a cytotoxic agent within a certain time frame to being able to divide effectively indefinitely. To study this aspect of cell survival, compromises have to be made so that one can define a series of criteria that can be used in an objective fashion. Such compromises are common in scientific experimentation.

The convention that has been adopted here is that cells are said to survive after a particular treatment (i.e., be alive), if they are capable of undergoing about six cell divisions or are capable of producing a minimum of fifty daughter cells within a prescribed time limit. An obvious compromise here is that six divisions should in theory give sixty-four cells, if each one divided on every occasion. For convenience, this is commonly reduced to the figure of fifty daughter cells produced. These studies are easy to perform using cell culture systems, where each daughter cell may remain attached to each other and form a *colony*, which because they are derived from a single surviving cell are also referred to as *clones*. The single cells that give rise to the colony are called *colony forming cells* or *clonogenic cells*. Clones here have a slightly different connotation from the current debate on cloning animals such as sheep, mice, and humans, where the animal is being derived from a single manipulated cell (as discussed), but here it is required to generate a diversity of differentiated tissues and many billions of cells.

These experimental approaches are essentially designed to study cells that survive a particular treatment by the maintenance of their reproductive potential and the cells that die, by definition, are those that lose their reproductive potential or become sterilised. One issue here is that, within limits of a set time frame, some cells may not satisfy the criteria because they are reproducing too slowly but if a longer time frame was used they also would be capable of producing fifty cells. Another complexity is that the cells may not always divide in a truly exponential fashion, that is, one cell producing two, two producing four, and so on, but the tissue or colony may be organised in a more complex fashion with cells entering differentiation and different categories of proliferating or reproducing cells (cell lineages or cell hierarchies; see Chapters 6 & 7).

Cell death: human analogies

So, if metabolism, movement, sensory perception, and reproduction are the basic features of life, death must be characterised by the absence of these features. Is it necessary for all these processes to cease for death to be defined? Cells may stop movement but still be capable of metabolism; they may stop reproduction and still be capable of metabolism and movement; they may stop reproduction and movement but still be metabolically active. The difficulty one encounters here is very similar to the modern ethical and philosophical problems associated with defining the point of death for a human. It might be worthwhile to digress for a moment and consider these questions in somewhat greater detail.

If an individual capable of movement suddenly ceases to move, a concern might arise as to whether death has occurred; however, the individual may simply be asleep (from a cellular point of view, in a dormant state). Careful examination would reveal a persistence of autonomic, involuntary movement, for example, the chest moving as a consequence of breathing, evidence of blood circulation, and eye movements. The presence of such movements somewhat alleviates the immediate anxiety as to whether death has occurred. However, the individual could be in a coma; now the situation is more serious. Breathing still occurs, movement of the blood around the vascular system continues, and the tests for whether life exists become somewhat more stringent. Does the individual respond to sensory perception? For example, does a person wake up if he or she is pricked with a needle? Is the main sensory organ, the brain, functioning properly? In fact, different levels of brain function may need to be determined: cerebral activity, brain stem function, and so on. Such situations also inevitably lead one to address the following question: 'Can metabolic activity still be demonstrated?' Obvious food intake may have ceased for the individual but questions arise as to whether the person is still taking up oxygen (breathing), still excreting waste (full kidney function), and so on.

Failure of any one of the major organs of the body can precipitate death. The kidneys may fail and the levels of blood waste products, most noticeably urea, will rise. The liver may fail and because it is a major metabolic organ many other toxins can be built up in the blood. The lungs may fail and the blood gas levels will change. All of these are easily detected by appropriate blood tests but if not dealt with can prove fatal. One can suffer an aneurysm of the main aorta as a consequence of cholesterol buildup; effectively the aorta bursts. There can be various defects associated with the heart, ranging from ventricular fibrillation (where failure to pump blood properly results in the brain not receiving enough oxygen) to coronary thrombosis or myocardial infarction, which are essentially characterised by a blocking of the blood supply to various regions of the heart, producing the symptoms known as a heart attack. Some of these defects can be detected by checking the pulse, using a stethoscope or, ultimately, using an electrocardiogram (ECG), which analyses the contraction of the heart. Finally, there are various problems associated with the brain and the brain stem that can result in rupture of blood vessels from head injury, or from infarction caused by clots in the blood vessels supplying the brain. Both these events result in starving the brain first of blood, *ischaemia*, and, consequently, oxygen (*hypoxia*). The severity of brain injury depends on the size and the location of the ischaemic area and the consequences are the symptoms known as a stroke. There are a whole variety of levels of unconsciousness or coma that can occur that are generally tested for by looking for electrical activity in the brain using an electroencephalogram (EEG). The most severe state would give little or no reading on an EEG for the brain cortex, but if there is still brain stem function then breathing and heart activity and one or two basic reflexes will continue to be detected. However, such a person would be in a persistent vegetative state and would require life support systems.

Now, a feature of all these defects in the human body is that, provided that the patient can receive adequate medical treatment and can fairly rapidly be taken to hospital, many of them can be overcome in modern medical practice by the use of appropriate equipment: kidney dialysis machines, heart-and-lung machines, and so on. So although these individuals would have undoubtedly died if they had not received appropriate treatment, they can be rescued from death by modern techniques. People on dialysis machines, those fitted with heart pacemakers, and so on, can lead a relatively normal life. The most difficult cases are those in coma, particularly those in a persistent vegetative state, and the question to what extent such individuals can be regarded as alive remains a matter of some argument. Once brain stem activity is lost, although a patient can be kept technically "undead" on life support systems, switching off these systems will inevitably lead directly to death. Major legal and ethical questions arise in these cases. The difficulties outlined here can also be considered in relation to single cells in tissues or in a culture dish.

In modern medicine the ethical problems arise because a person can lose many of the key functions of life and yet be kept "alive" by appropriate machines. The kidneys may have failed, the lungs may have ceased functioning, and the heart may even have stopped pumping oxygen around the body, but the brain may still be capable of some function and the individual cells of the body are clearly still alive. The individual cells are still metabolically and reproductively active and will function normally if transplanted into another individual. So, at what point does one define death in this sort of situation?

One or two real-life examples may reinforce the difficulties to which I allude. At the 1989 Hillsborough football stadium disaster in the United Kingdom, many people were crushed and a few were unconscious when taken into hospital, where they were kept in a persistent vegetative state by intensive maintenance theory. Eventually the courts ruled that in one case support systems could be switched off, that is, the final stages of dying were permitted to occur. However, another patient from the same accident in a similar vegetative state continued to receive supportive treatment and eventually was able to communicate again, at least to a limited extent.

An often-quoted case concerns a man from Minnesota involved in a serious car accident in California in 1978. He required several blood transfusions after the accident. He remained unconscious and could not maintain unaided respiration. No next of kin could be traced and a transplant team was notified because his heart and kidneys were undamaged. Appropriate consultations took place and in California he was legally pronounced brain dead. However, some time later a son arrived and legal action was taken, based on the fact that in Minnesota brain death was not recognised. So in one state in the United States he was legally defined as dead, whereas in another he was legally defined as alive at the time the ventilator was switched off and his organs were removed. Such cases illustrate the difficulties, both legal and ethical, in defining death.

Similar complications arise when we define death as loss of reproductive potential. Clearly for adult humans, the question of whether they can still reproduce is not generally considered part of the definition of life or death. Women beyond the age of menopause are not categorised as dead! But for cells this is a fairly common criterion by which their full functional competence is assessed, and the loss of reproductive ability is regarded as a serious marker of loss of cellular functional activity and hence cell death.

These considerations apply also to individual cells, whether they are free living or part of the tissues of the body, and one is forced to consider the following question: 'At what point does a cell die and loss of which functions are most appropriate to use to define the loss of life?' Cells can clearly enter phases of dormancy or inactivation during which most of the functions of life might be regarded as being absent (extreme examples are seeds or spores). These cells can be brought back to life after even long periods and harsh environmental conditions by providing the right signals or environment. Cells can become quiescent in tissues of the body when they do not continue to progress through the sequential metabolic processes leading to cell division. They may slow down or stop their movement, relative not only to other cells in the tissue but also to the internal movement of the cytoplasm. They might become much more resistant to various changes in the environment (loss of sensory responses) and certainly they may cease to reproduce.

It is therefore important to remember that death is not as simple a state to define as we might at first imagine. Of course, once true death has occurred it may be very easy to recognise and we discuss this further when we consider the appearance of dying or dead cells. Again, it is necessary to think of the human situation – only when it becomes very obvious that once death has really occurred is it easy to recognise. The body starts to dehydrate, decay, and disintegrate.

Cell death is often characterised differently depending on the scientific interests of those involved. To the biochemist, life may be defined simply by the presence of metabolic activity, whereas to those working on cancer drugs it may be defined by a cell's ability to divide not only once but many times.

Another interesting consideration arises from some insects and higher creatures such as the North American wood frog, which can be frozen solid in winter and by almost all criteria would be defined as dead. However, with the advent of spring the animals thaw and the heart is kick-started into pumping and the recognisable signs of life return.

When looking at the frozen-solid North American wood frog, it is difficult not to be convinced that the animal is dead. No signs of life can be defined using the criteria oulined earlier. To reach some conclusions about its status one has to know something about its past and/or its future; that is, was it alive or dead when it was frozen? Also, will it be alive or dead when thawed? It is clearly 'undead' when frozen.

Another amazing example of animals appearing to be dead but are able to come back to life is given by little animals called *Tardigrades*. These creatures are approximately 0.05–1.2 mm long and live in the moisture layers around soil particles. There are many species and when viewed through a microscope they have a distinctly bearlike appearance, with four legs with clawlike bristles and a very bearlike face; as a consequence they are called water-bears. The surprising fact about water-bears is that they can reduce their body water content from 85 to 3 percent; that is, they can dehydrate and then be resuscitated by the addition of water (a process called cryptobiosis). They can remain dehydrated for many years and in this state withstand extremes of temperature and also large doses of radiation. If one was found in its dehydrated state, a person would conclude that it was dead because it exhibits none of the signs of life; however, exposure to water brings it back to life. Some believe such creatures could traverse interplanetary space on meteorites and may have arrived on earth that way. There are several other species of worms and insects that can survive freezing.

Another example of surviving freezing can be found in the research laboratory. Most mammalian cells used in cell culture studies are routinely frozen in liquid nitrogen for long-term storage. On removal from storage, the cells can be thawed and placed in culture medium in a warm incubator and will restart their metabolic processes and proliferate. Sperm and eggs are also routinely frozen prior to use for *in vitro* fertilisation. In some cases people who suffer cardiac arrest can be resuscitated: The heart has stopped, brain function and other signs of life are absent, but application of electrical stimulation to the heart may result in the person being brought back to life. So, the distinction between life and death can sometimes be difficult to make; or, to be more specific, "death" may not be as permanent or irreversible as it has always been thought to be.

It has been proposed that all the cells of our body are actually programmed to die but are kept alive by a battery of signals and messages instructing them to survive, proliferate, differentiate, and perform their functions. These signals come from a variety of sources: the molecular superstructure on which the cells sit (the extracellular matrix and the basement membrane), the cells in the supporting connective tissue, and neighbouring cells in the tissue in question and from cells at distant locations in the body. These signals may be derived from physical contact or from secreted factors (e.g., polypeptide growth factors and hormones). As stated above, some of these signals are crucial survival signals that prevent the cell initiating the programmed series of events that result in apoptosis. During the course of evolution and development of complex multicellular organisms considerable redundancy has evolved so that if the primary signal is damaged, deleted, or altered by mutation, other signals and systems swing into action to compensate for the loss. When all of these signals are used up or absent or if other overriding messages are activated, the cell may undergo its predetermined death.

Consequently, the philosophical question that arises is not so much 'Who am I?' or 'Why am I here?' but rather 'Why am I not dead?' And it becomes more important to consider the precise definition of death rather than the definition of life, which could then of course be simply 'not dead'.

We have already touched on the difficulties of defining the moment of death for individuals and the fact that there are similar difficulties that are encountered when considering individual cells. An interesting example to consider for a moment are the cell type called fibroblasts, which are found in the supportive connective tissues of the body. This tissue has been called the mesenchyme and fibroblasts are mesenchymal cells. In the body they are responsible for producing molecules that provide strength and elasticity to the tissue (by producing fibrous molecules such as collagen and elastin, amongst others). In the body, they are characterised by the fact that they are largely nonproliferative (i.e., they do not divide). Curiously, these cells are extremely easy to grow in culture and are often annoying contaminants when other cell types are being prepared for culture. In the culture environment, quite unlike the situation in the animal, fibroblasts proliferate rapidly and with ease. Because of their ease of growth, these cells have been used to study a wide variety of basic biological processes associated with cell division and apoptosis. It is perhaps interesting that in both these processes their behaviour in culture is much different from that in the tissue. This is a problem associated with many cell types and in fact epithelial cells from the epidermis in the skin can also be grown in culture, but with time and repeated divisions they gradually lose the characteristics that define them as epithelial in origin and may assume characteristics typical of fibroblasts. Thus the use of cell culture systems to study certain fundamental aspects of the behaviour of cells in tissues has to be considered with some reservations.

Interestingly, from the point of view of intercellular communication, the epithelial cells from the skin (keratinocytes) grow well only in the presence of fibroblasts (commonly referred to as "feeder cells"). The feeder cells are generally irradiated with very high doses of radiation to stop them dividing, but the cells remain intact (i.e., are metabolically active) and continue to produce the epithelial survival and growth signals.

Fibroblasts in culture can be induced into a vegetative or quiescent state somewhat similar to that of a patient in a vegetative coma. This can be achieved by growing the cells until they fill the culture dish (the cells become confluent) or by altering the culture medium or culture environment (e.g., by removing certain factors from the medium). This can also be achieved, rather surprisingly, for example, by irradiating the cells with X-rays. These so-called lethally irradiated fibroblasts do not die in the morphological sense; that is, they do not change their appearance and become obviously necrotic (swell and rupture) or apoptotic (condense and fragment). Rather, they remain static for protracted periods of time in a state similar to quiescence but different in that they cannot be stimulated to proliferate. In this respect, they are again behaving in a manner that differs from cells in the tissues of the body. Are they dead or alive? Is a red blood cell (erythrocyte) dead or alive? It exists as a mature functional cell for about 40 to 45 days in a mouse, or 115 days in a human, performing its function of transporting oxygen round the body. It actually lacks a nucleus in mammals, which it discards at the last stage in its maturation from an erythroblast, and consequently lacks any genetic regulatory activity while performing its differentiated function.

Thus all cells may have the programme for death present and it may even be activated. However, they are prevented from executing this programme in most cases by signals from the surrounding environment. If it is desirable, the activated cell death programme may be allowed to run its course. This may occur from stages as early as the fertilised zygote. We know that up to 70 percent of human conceptions abort at a very early stage and this involves apoptotic cell death and that this is apparently regulated by certain genes whose job it is to monitor the genetic integrity of the embryonic cells. If these genes are deleted, many of these spontaneous abortions are prevented and many foetuses develop with congenital abnormalities.

2

How to die?

With there being several criteria on which a definition of life or death can be based, it is then difficult to give a precise definition of what is a living cell and what is a dead cell and the precise transition point at which life becomes death. As stated previously, cells dying by apoptosis or necrosis change their appearance in a readily identifiable way and this is an often-used criteria to define cell death in biology. In this chapter we consider more closely what the decaying corpse of a cell actually looks like, but first let us consider those situations where a cell dies and perhaps does not display all the morphological changes that we associate with this death.

The undead

There are perhaps two circumstances in which cells may have passed part of the way down the death pathway and yet are to all intents and purposes indistinguishable from other normal healthy looking cells. The first of these might occur in a situation where damage has been induced in a cell resulting in the initiation of some of the steps towards cell death, or where a healthy undamaged cell has received some external signal that it should die, but completion of the sequence of events required to change its appearance from living to dead has been prevented. It may have received a second signal, or message, telling it to stop the process (in other words, a counteracting survival signal), or it may find itself in a situation where it can no longer undergo the metabolic processes required for this transformation from a living appearance. What I mean here is that it can no longer undergo the DNA transcription, translation, and protein synthesis that seems to be needed to complete this programmed change from living to dead appearance. The cell may have found itself in an environment that prevents these metabolic changes, the temperature may have dropped, or the supply of vital nutrients or substrates may be absent and consequently the cell essentially goes into a state of rest or dormancy. This is somewhat analogous to the situation of the frozen corpse in a morgue, which does not alter its appearance with time (i.e., does not decay), but is dead in this case and cannot be resurrected or, as mentioned in Chapter 1, the North American wood frog that freezes in winter but is 'resurrected' by the advent of spring.

There are some who would argue, I think convincingly, that many or most of the cells in our bodies have the programme for death present and that this programme is potentially active. The cells are prevented from expressing this programme by a barrage of signals that tell the cell to survive and function. If this is the case it is easy to see how, as a cell approaches the end of its life, the levels of these survival signals may be weaker or the signals may be less well received and death of cells then ensues.

A related situation to consider is the predominant number of cells of the body that are nonreproducing cells performing their function, a process that we refer to as differentiation and the cells as differentiated, functional, or mature. The examples are numerous: the vast numbers of cells in the upper layer of the skin, full of their differentiated product keratin, the absorptive cells lining the intestine (cells secreting mucus, or enzymes in the intestine, or hydrochloric acid in the stomach), red blood cells, some white blood cells, and of course the specialised cells in the brain and muscles. Each of these cell types is characterised by producing specialised and different proteins and, on the whole, such cells completely lack the ability to divide. These cells are produced in vast numbers by the reproductive, or proliferative, cells of the tissue and live a variable length of time in the body ranging from several days through to perhaps the life span of the individual.

Whether these cells can be defined as dead is debatable. Are they analogous to postmenopausal women – knowledgeable in the ways of life and highly functional but nonreproductive? Such cells are dead in terms of loss of reproductive potential, and they may be dead within the framework of the hypothesis just outlined, which suggests that all cells are programmed to die. In this connection, there is certainly some evidence that differentiated cells in the intestine, as they near the end of their short functional life of a few days, express more of the characteristics and features associated with programmed cell death, specifically showing some of the metabolic changes and gene expression that precede the changes in appearance. Similarly, the cells in the upper layer of the epidermis in skin (the granular layer) condense their DNA or chromatin prior to degrading the DNA and nucleus – steps that are also common to apoptosis.

The clearly dead

In a variety of situations, for example, during development in normal healthy tissues, the development of tumours, and after exposure to damaging agents, cells with a changed morphology can be recognised. The appearance of these abnormal-looking cells was detailed very early in the days of microscopy and a variety of names have been given to these altered cells at various times in the past. Since the mid-1970s, the picture has been made somewhat clearer by the use of the electron microscope and much more sophisticated and powerful staining methods for sections of tissue for viewing in the light microscope. A pioneering article was produced in 1972 by an Australian pathologist, John Kerr, and two pathologists initially from Aberdeen, Alistair Currie and Andrew Wyllie. They pointed out that there were two distinct types of dying cells characterised by different appearances and one of these was associated with an element of self-regulation and control by the cell, that is, is programmed in some way. They were the first to apply the term *apoptosis* to this controlled cell death and to identify it as an important physiological and pathological process that was distinct from necrosis (table 1 and figure 1).

TABLE 1	
Apoptosis	**Necrosis**
General	
Affects isolated cells	Affects clusters in cells
No inflammatory response	Inflammatory cells invade tissue
Metabolic	
Early increase in protein & RNA synthesis	Switching off protein synthesis and phosphorylation
Cell membrane remains intact	Cell membrane becomes leaky
Affected by protein & RNA synthesis inhibitors	
Endonuclease activation (calcium- & magnesium-dependent)	
ATP-dependent	ATP-depleted
Morphological	
Loss of cell to cell contact, early	Loss of cell contact, late
Shrinkage of cell	Swelling of cell (sodium ions and water taken up by cell)
Membranes remain intact (no enzyme leakage)	Membrane defects (internal & external) (enzymes leak out)
Organelle structure maintained, late	Organelle structure lost, early
Condensation of cytoplasm	Cytoplasm becomes clear
	Mitochondria swell and acquire dense bodies.
Condensation and margination of chromatin	Lysosomes disrupted (release of enzymes)
Sharp edges to chromatin masses	Irregular edges to chromatin masses.
Nuclear fragmentation	No fragmentation
Cytoplasmic budding and fragmentation	
Phagocytic removal	

Necrosis

This type of death tends to be associated, largely but not exclusively, with rather severe forms of cellular damage. One can think of it somewhat like the "being hit by an express train" or "run over by a bus" type of death, where the individual has little time to make any decisions, arrangements, or do anything about his/her demise. From

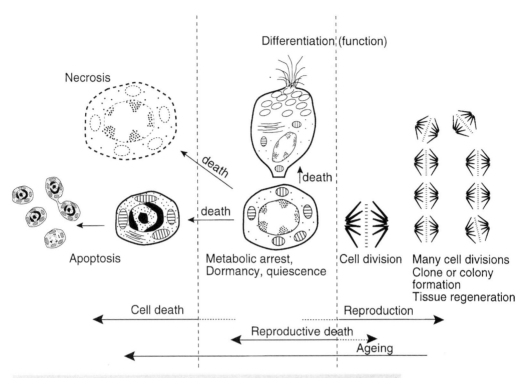

1. Diagrammatic representation of cell fate, showing commonly observed morphological features.

a cellular point of view, this type of death is induced by physical disruption or severe metabolic poisons. The death is often characterised by clumps of cells in a tissue acting together where generally speaking the cell swells and the cytoplasmic organelles start to disintegrate very rapidly. There is no evidence of continued metabolic activity. If one imagines heating the cell, the protein and cytoplasm coagulates (as in a boiled egg), and in fact this form of death has sometimes been referred to as *coagulative necrosis*. The DNA in the nucleus condenses, particularly at the margins, in an irregular fashion and the DNA and cellular constituents start to disintegrate in a random and uncontrolled fashion. Vital internal constituents rapidly leak out of the cell and the host tissue recognises this and floods the system with inflammatory lymphocytes to stop infection and damage to the tissue being induced (an inflammatory reaction). The cellular debris is

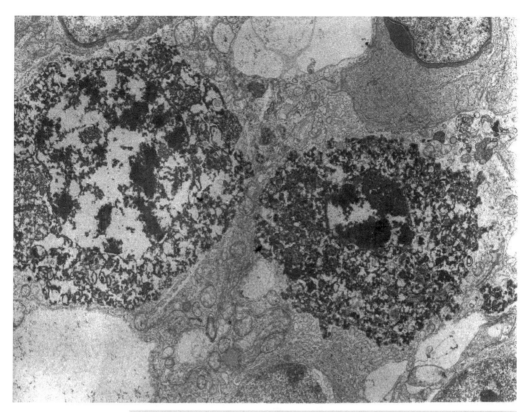

2. Electron micrograph of necrotic cells. Two cells may be observed showing characteristic nuclear and cytoplasmic degradation (courtesy of Dr. D. Allan, Brisbane).

then engulfed and removed by specialised cells called macrophages. Figure 2 shows what such a necrotic cell looks like in the electron microscope.

Apoptosis

The second process of death was initially referred to as *shrinkage necrosis* and subsequently renamed *apoptosis*, a term that has led to some conflict in the scientific community in terms of its precise meaning, derivation, and pronunciation. The term was suggested by the Professor of Greek at Aberdeen University, James Cormack, and in Greek it means the falling away of leaves of deciduous trees, petals from flowers, or cells from a tissue. A similar term was already in use to describe the drooping of the upper eyelids. The second 'p' in apoptosis is generally silent although the pronunciation inevitably varies. As suggested by the foregoing, apoptosis is characterised by an initial shrinkage of the cell, which in a tissue means breaking cell-to-cell contacts with neighbours and rounding up. The consequence of this stage tends to be that the volume of the cell becomes smaller, and the cytoplasmic internal membranes, ribosomes, mitochondria, and other cytoplasmic organelles are more concentrated in the cytoplasm, which then consequently looks darker. The organelles remain intact and healthy looking very late into the process of death, suggesting that the cell continues metabolic activity for some considerable time. This indeed can be demonstrated in many cell systems by using various inhibitors of metabolic processes (RNA and protein synthesis inhibitors), which will delay the progression of a cell through the apoptotic sequence. Apoptosis characteristically involves single isolated cells and not clusters. Because the membranes remain intact, cytoplasmic constituents do not leak from the cell, and an inflammatory response is not mounted. The DNA or chromatin material in the nucleus condenses, again at the margins of the nucleus, but in the case of apoptosis this condensation is very extreme and results in areas of condensation with very sharp boundaries, commonly generating crescent shaped areas of condensed chromatin that follow the contour of the nuclear membrane. This used to be referred to as *pycnosis*. Concomitant with this is a process of cleaving the DNA into fragments. This process involves a very specific enzyme called an *endonuclease*, which characteristically breaks the DNA between the clumps of chromatin that are referred to as nucleosomes. The consequence of this is that the DNA breaks into fragments of rather precise sizes that are multiples of about 200 DNA base pairs in length (180 base pairs to be precise). Whilst these processes of DNA cleavage are going on, the nucleus begins to break into fragments and the cell likewise splits into pieces; thus the original dying cell generates a number of cellular fragments, *apoptotic bodies* or *apoptotic fragments*. This used to be referred to as *karyohexis* or *karyolysis* (see in this chapter). Each of these fragments may, or may not, contain a fragment of the original nucleus and hence some DNA. Finally, these fragments

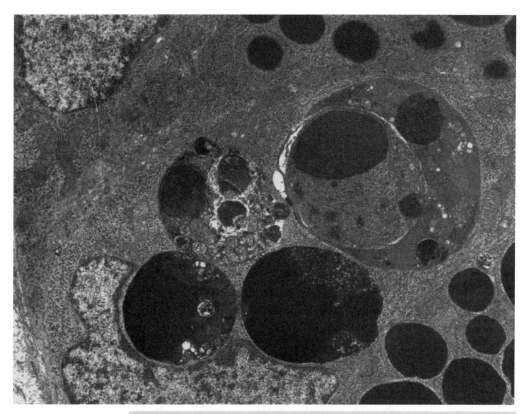

3. Electron micrograph showing a healthy cell in the small intestinal epithelium that has, over a period of time, engulfed apoptotic fragments from at least four neighbouring cell death events. Indeed, one of the dying cells (the upper fragment) had previously engulfed a fragment from yet another dying cell. These fragments show different levels of digestion. The healthy cell nucleus is in the lower right. The large black granules are the secretory granules in neighbouring cells (called Paneth cells) (courtesy of Dr. T. Allen, Manchester). Reproduced from Merritt et al., 1997, with permission.

are engulfed or eaten by the specialised migrating macrophages or healthy neighbouring epithelial cells by a process called *phagocytosis*. This phagocytic process results in apoptotic bodies enclosed in a membrane-bound vesicle in a cell called a *phagosome*. The host cell then gradually digests the apoptotic body. One of the features of apoptosis is that the cytoplasmic organelles may continue to look healthy and indeed may be functioning even when they are enclosed in a phagosome in a host cell. Figure 3 shows what this looks like at the level of the electron microscope and figure 4 shows a diagrammatic

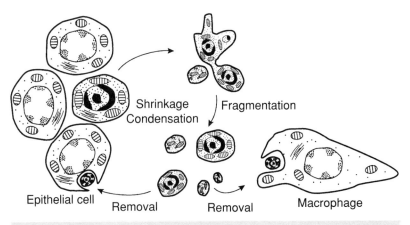

4. Schematic representation of the apoptotic sequence.

representation. Ultimately the phagosome and its contents are degraded, probably to a body referred to by electron microscopists as a *myelin whorl* (the indigestable remains of the corpse).

Situations where death might be initiated

We have already discussed situations that might lead to the induction of necrotic cells. This can be thought of as a form of *cell murder* or *cell slaughter* induced by rather extreme forms of cellular damage.

When we consider apoptosis the situation is more complex. Apoptosis is a form of cell suicide involving some controlled or programmed processes within individual isolated cells. There are essentially three situations under which one might expect to see apoptosis (table 2). The first of these we can regard as *utilitarian cell suicide*, that is, the death of one cell for the good of the majority.

This is a situation where perfectly healthy cells are instructed by the environment in which they are located to delete themselves or commit suicide because too many cells are present, or space is required for a repositioning and reorganising of the tissue. This occurs extensively during developmental processes in the embryo and this situation is discussed in greater detail in Chapter 8. Malfunctions in developmental utilitarian cell suicide result in deformities and malformations of the embryo. This type of apoptosis may persist in at

TABLE 2	
Necrosis	Apoptosis
Cell 'murder'	Cell 'suicide'
Groups of cells	Single cells
Cell slaughter ('hit by a bus')	Altruistic cell suicide (removal of defective cells)
Heat, poisons	Utilitarian cell suicide (for the good of the greatest number)
	Senescent cell suicide (deletion of old 'worn-out' cells)

least some adult tissues, as part of the process whereby cell numbers are regulated and occasional healthy cells may be deleted to ensure that a structure in a tissue, or the tissue as a whole, maintains its correct size. The important feature of this type of cell death is that it involves the removal of perfectly healthy cells; there is no damage induced, no damage recognition, and therefore no utilisation of damage recognition/damage response mechanisms.

The second situation, which involves the induction of apoptosis in undamaged cells, is apoptosis that is associated with senescent, worn out, or aged cells. We could regard this as a process of *senescent cell suicide*. It is not entirely clear how widespread this process is; however, there are some good indications that cells in the intestine switch on some of the initial steps involved in apoptosis as they reach the end of their normal life expectancy in the tissue, which may be three to seven days after they were born, by cell division in the proliferative compartment. There are some who would argue that cells in the upper layers of the epidermis, or even erythroblasts when they become erythrocytes, also exhibit some changes consistent with having initiated apoptosis, although this is much more debatable.

From a theoretical point of view, there are two possibilities here. The first is that the cells themselves recognise that they are at the end of their life and initiate the suicide programme. The second possibility is that the tissue, or other cells, recognise that a particular cell is at the end of its life span and instruct it to initiate apoptosis. We can think of this perhaps as *assisted or facilitated cell euthanasia*.

It is not at all clear which of these two possibilities actually occurs. It is also possible that as the cells become old they accumulate DNA errors and that what we see here is an extension of the error removal type of apoptosis (see below). With the progressive accumulation of errors the cells become closer and closer to the threshold that triggers the damage-induced apoptotic sequence.

The final situation under which apoptosis may be triggered in a tissue is as a consequence of damage to the cell. This would be damage of a type and level of severity that does not result in necrosis but is consistent with the continued metabolic activities required for apoptosis. This might be considered *altruistic cell suicide*. Again, we have a theoretical distinction here as to whether the cell itself recognises that it is damaged and, without any external aid, initiates the process of apoptosis. The theoretical alternative to this is that the tissue or neighbouring cells recognise the damage and instruct a particular cell to commit suicide. The evidence would favour the concept of altruistic cell suicide, mainly because some of the genes that are involved in the damage recognition response process have been identified, which implies that the cell is undergoing a damage self-screening process. The implications here are that the cells can detect even low levels of damage to vital structures such as DNA or other vital elements such as membranes and, rather than attempt to repair this damage, they commit suicide. Repair itself may be associated with a certain risk of making a mistake and a mistake in the DNA could result in a disastrous mutation. This type of mechanism, which has been clearly demonstrated to operate amongst important reproductive cells, or proliferative cells in the intestine, ensures that there is little chance of DNA damage being perpetuated in the form of a genetic mutation and, therefore, also in the form of a potential carcinogenic change.

The appearance of cells as they undergo the process of apoptosis has been summarised earlier (table 1); however, in some cellular systems, most notably cell culture systems, the dying cells do not necessarily follow the same sequence through to the final fragmentation. In some situations, the chromatin condensation results in a dense spherical ball of chromatin or a single mass of chromatin at one end of the original nucleus. In some situations the cell remains a single apoptotic body. Hence, electron microscopic and light microscopic pictures presented in the literature can sometimes look somewhat different from one system to another.

In tissues and tumours, the apoptotic fragments tend to be removed from the tissue by infiltrating macrophages or local neighbouring cells. Once phagocytosed, they are 'digested' by the host cell and so gradually undergo further degenerative processes. The chromatin mass tends to lose any morphological association with nuclear structure and the cytoplasmic organelles begin to disintegrate. These late stages of apoptosis can somewhat resemble the electron microscopic appearance of cells undergoing necrosis. It is believed that the digestion by the host cell of the apoptotic fragment continues, ultimately generating in the cytoplasm of the host cell, a small structure consisting of concentric lamellae (membranes) called a myelin whorl, which can be rather loosely thought of as the indigestible remains of the host cell. However, a word of caution should be inserted here. This sequence of events is based on the interpretation of individual pictures that represent an instant in time for a particular cell. It is an unfortunate necessity that dynamic processes that cannot be viewed directly in real time have to be interpreted from a series of static snapshots and, consequently, there is always the possibility that events in the sequence are missed or misinterpreted.

So, bearing all these points in mind, published photomicrographs of apoptotic cells that appear in numerous articles and texts vary somewhat in appearance. This is due to the fact that different cellular systems are involved and the pictures tend to be generated using somewhat different approaches (fixations, treatments, etc.), are of varying quality, and may represent different stages in a dynamic process.

Attempts have been made to record the sequence of events, using time-lapse videomicroscopy, for various cells grown in culture. On the whole such culture systems, although they mimic the situation in tissues in some aspects, differ in a number of other major respects. The major differences are that the cells do not possess the same cell-to-cell and cell-basement membrane connections and are not exposed to the same range of growth factors and hormones that may be found in living tissues. The absence of these external signals and stimuli affect the cell in relation to all its activities (i.e., metabolism, division, death, and differentiated state). Despite these caveats, time-lapse studies do show that the process of apoptosis is a very dynamic one under these conditions. The cells initially tend to round up and break their attachments to the plastic or glass surface on which they are growing and to any neighbouring cells. They then undergo a series of violent and rapid contortions, throwing out protuberances, or blebs, from their surface and then drawing them in again. This sequence of death throes, called by some the *"dance of death,"* can continue for some time and is probably the culture equivalent of the fragmentation process that we see in tissues. However, in culture it is not necessarily completed to generate a series of apoptotic fragments. These violent contortions again demonstrate that the process is a very active one for the cell and considerable energy resources are required to execute the contortions, which must also involve the internal skeletal elements of the cell, the filaments of actin and other proteins, such as tubulin and myosin, in some cells, and the intermediate filaments, which are smaller structures, such as the cytokeratins in epithelial cells and vimentin in many of the other cell systems. Some of these proteins are contractile and so may account for the contortions that are seen.

It is clear, therefore, that a cell undergoing apoptosis can generate a number of apoptotic fragments, from, at the one extreme, a single fragment that represents the entire dying cell through to as many as a dozen small fragments, which together represent the volume that was the original cell. It is not at all clear what determines the degree of fragmentation, other than to say that in many *in vitro* (cell culture) systems the dying cell tends to generate one apoptotic body or apoptotic cell, whereas *in vivo* (in living tissues) many fragments are generated. Even in one tissue following one apoptosis-inducing protocol, some cells may die and generate only a single fragment, whereas others may generate up to twelve fragments (figure 5). In this particular case this represents the number of closely associated fragments observed in the proliferative units of the intestine (called crypts) following a small dose of radiation. The measurements are

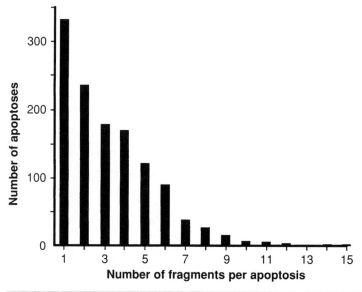

5. A frequency distribution for the number of fragments (apoptotic bodies) generated from individual dying cells in the small intestine following a dose of radiation, determined by observing apoptosis in whole pieces of tissue. Reproduced with permission from Potten, 1996.

made by observing the clusters of fragments seen in entire crypts dissected from the tissue placed on a microscope slide and carefully observed by focusing throughout the entire structure. There is some suggestion that the number of fragments generated may, to some extent, be determined by the severity of the cytotoxic insult although this has not been conclusively proven.

How long to die?

The question of how long it takes for a cell to go through the sequence of events described above remains unclear although the scientific literature contains somewhat dogmatic statements on this subject. The problem here is that the length of time that it takes for the process to occur depends on the efficiency with which one can recognise when it starts and when it finishes, and what techniques one is using to identify the process of apoptosis, and which cellular system is being studied. The duration almost certainly varies considerably depending on whether one is considering cell death in invertebrates, in early mammalian development, in later mammalian development, in mammalian cells grown in culture dishes, or in mammalian cells within the framework of a highly organised tissue. Some observations based on studies in culture suggest the process is extremely quick (over a time frame measured in tens of minutes). Here the observations may be based on simple stages in the process such as the detachment of cells and their loss from the field of view (off the culture dish into the culture medium) or the number of cells that have reached a recognisable, specific morphological end point (i.e., shrinkage and condensation, with regard to apoptosis) at any point in time. Intuitively, one would imagine that the apoptosis that occurs in developing embryos must occur quickly because all other processes here also occur quickly; the average time between successive divisions can be very short and the structure itself is growing very rapidly.

Attempts have been made to measure the time taken for apoptotic cells to disappear in a tissue. In the case of the epithelial lining of the small intestine, the duration of apoptosis is somewhere between three and twelve hours. In this tissue the cells divide on average twice a day and there is extensive movement of cells within the tissue, as a consequence of all the division activity. (This tissue is described in greater deal later because it forms a model system in which we can study various aspects of apoptosis *in vivo*.) These *in vivo* studies suggest that the duration of apoptosis is of the same order of magnitude as the average intermitotic time or the time it takes to replicate the DNA.

Occurrence of apoptosis

It has become clear since the mid-1970s that apoptosis, or programmed cell death, is a fairly universal phenomenon. It occurs in a precisely regulated fashion during the growth of invertebrates and has been extensively studied in a small microscopic roundworm (belonging to a group of organisms called nematodes). This worm has the name *Caenorhabdites elegans* (*C. elegans* for short – see Chapter 5).

The adult worm contains exactly 1090 cells and during the growth of the worm from a single fertilised zygote there are clearly a large number of cell divisions but there are also exactly 131 cell deaths. These occur at defined times and in specific cells and any deviation from the pattern results in malformations of the adult worm. This is probably the most extensively studied system but by implication similar processes are occurring in all other invertebrate and vertebrate systems, including the embryos of higher mammals such as humans. Apoptosis appears to be an important process in the very early stages of embryo development in removing embryos that contain abnormalities. Large numbers of human and mouse fertilisations effectively abort at very early stages as a consequence of apoptosis. However, apoptosis is also important at later stages when, as the embryo develops, changes in the shape and size of tissues require the removal of many cells to allow tissue restructuring: Cells must be removed to allow space for other cells to develop. There are many examples where this is important; one is during the formation of the palate in the mouth and another is the formation of fingers and toes in humans. In the embryo there is webbing between the fingers and the toes and this interdigital web has to be removed and it is removed by apoptotic cell death in the web tissue. So as a simple example, if it were not for apoptosis during embryo development humans would have webbed hands and feet!

There are many other situations where apoptosis plays a similarly important role. It is important in insects during metamorphosis, during the transition between caterpillar and butterfly, and in the development of amphibians and reptiles. Another classic example is the removal of the tail of a tadpole as it converts into the tailless adult frog. Here, the tail is resorbed gradually by the animal and this involves the induction of apoptosis in healthy tail cells. To initiate specific cell death events during development, apoptotic signals within the embryo must be generated at the appropriate time. These signals may be in the form of the synthesis of death-inducing molecules or the failure to provide vital life-maintaining signals to a cell. The production of these signals are under the control of special, regulatory genes. The cells to be deleted must have a mechanism to detect the signal (cell surface receptors for example) and a programme (defined biochemical pathways) internally that enables them to transduce the instruction carried by the signal and ultimately initiate the process of self-destruction or cell suicide. Of course, it is equally important that this signal is detected only by the right cells and stops at the right point that is, in a tadpole, the absorption process stops when its tail has disappeared and thus the frog does not start to absorb its own hindquarters! These examples illustrate the importance of apoptosis and the fact that it involves complex and precise programmes (see next section).

Apoptosis is therefore crucial in developing roundworms, insects, amphibia, mice, and humans. In most of these developmental systems the apoptosis is initiated in healthy, normal cells that are not required either at that moment in time or in the future of the developing embryo. As already explained, apoptosis is also important in removing cells that contain damage. This is important in the very early stages of development (discussed in Chapter 5) and this process involves slightly different elements of programming. In order for a cell to respond to damage it has sustained, particularly damage to the genetic material, the DNA, mechanisms must exist that enable the cell to recognise that its own DNA is damaged and this must involve some linear screening of the DNA helix. Once a defect in the DNA is recognised, the choice has to be made whether to repair the damage or to commit suicide in an altruistic fashion and remove the cell along with the DNA damage. In different circumstances these two alternative options have to be "considered" by the cell, that is, whether to repair or to commit suicide. Such decision making and the response to the recognition of internal DNA damage involve elements of genetic programming by the cell (see figure 6).

In adult mammals (especially mice and humans) apoptosis has also been fairly extensively studied. In adult tissues, unlike the situation in embryos, the size of the tissue is generally stable. Functional cells in the tissue, the so-called differentiated cells (discussed in detail in Chapter 7), become old and worn out and are shed. For example, red blood cells have a precise and limited life span. Cells from the surface of the skin are constantly being exfoliated (it is said that the

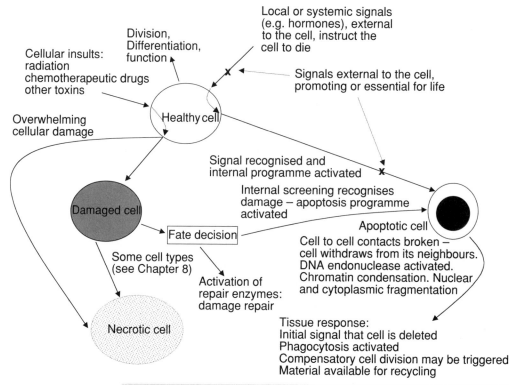

6. Cell programming and apoptosis.

bulk of the dust collected in a vacuum cleaner in a house is dead skin cells). Similarly, cells are constantly being shed from the surface that lines the intestine. These old, worn-out functional (differentiated) cells have to be constantly replaced and this is achieved by cell division in young reproductively active cells, located in specific positions within the tissue. For many years, it was assumed that cell division was precisely balanced only by cell differentiation. Cell differentiation can be regarded as another form of 'programmed' cell death. The cells are programmed to finish with reproduction and to make specialised products, perform the function for which those specialised products are needed and, when worn out, to die or be shed from the tissue. However, it is now clear that apoptosis is an equally important element in controlling the balance between cell division and cell loss. The regulation of the balance between cell production and cell loss is best achieved by having some cell deletion *via* apoptosis within the proliferative or reproductive compartment of tissue. Within the proliferative compartment cells divide and this division is carefully regulated by appropriate growth factors.

The number of cells produced by this cell reproduction must equal the number of cells required in the functional differentiated compartment of the tissue. If it does not, the tissue will either shrink or expand with the passage of time, and it has now become clear that part of the regulation of this process is that some cells are instructed to commit suicide, and if a proliferative cell commits suicide it removes not only itself but also all the daughters it might have produced in the future.

This regulation of the total cell production by controlling cell division, apoptosis, and differentiation is successfully employed in all the tissues of the body for the life of the organism (e.g., approximately seventy years for man or three years in a laboratory mouse). However, if the regulation is disrupted, tissues can shrink in size (as may occur with aging in humans) or expand as occurs in cancers. Thus apoptosis may be an important regulatory process that has great significance in our understanding of the changes that occur during ageing, but perhaps more importantly in our understanding of the development of cancer.

As in the early embryo, in adult tissues, particularly in the reproductive compartment, apoptosis appears to play an important role in removing cells that sustain and contain DNA damage. This process may be extremely important for removing cells that contain genetic (DNA) damage, particularly the DNA damage that may lead to cancer. The efficiency of this process, along with DNA repair mechanisms, contributes to the fact that cancer is an extremely rare event when one bears in mind the enormous numbers of cell divisions that occur in the lifetime of a human and the billions of cells that are produced.

What's in a name? A rose is a rose . . .

The term *apoptosis* was coined in 1972 by Kerr, Wyllie, and Currie in an article that appeared in the *British Journal of Cancer*. The importance of this article was not so much the coining of the phrase or the detailed description given of the process, but the recognition of the fact that the death of a cell was an important programmed process involving complex cell signalling systems, cell metabolism, and ultimately genetic elements (genes). The introduction of this term was initially strongly resisted by various scientific elements and scientific disciplines, and for at least ten to twelve years apoptosis research faced something of an uphill struggle. However, it was quickly realised by those working with developmental systems such as nematodes, *Drosophila* (the fruit fly), amphibia, birds, and mammals that this concept very well described the programmed cell death that was a common feature in the development of these systems. During the late 1970s and early 1980s apoptosis research continued primarily in the United Kingdom and Australia, and in the late 1980s flourished into a very rapidly expanding field with many groups in the United States and the rest of the world.

Inevitably, there has been some soul searching as to whether the Kerr, Wyllie, and Currie article was really the starting point of this research, but in our view it involved the formulation of the concept that cell death (apoptosis) involved an active biological process rather than the idea of a passive cell death process that tended to predominate prior to this date. It is clear, however, that by searching carefully through the literature, particularly the old German literature, observations on cell death were published as early as 1842. Studies on cell death in nervous tissue, cartilage, pupating flies, muscles, nerves, and in fish embryos were published in the periods between 1842 and 1906. Some of the descriptions in the latter part of the nineteenth century described events very similar to those that we now refer to as apoptosis. Cell death in breast cancer was reported as early as 1892 and the concept of apoptosis or cell death as a counter balance to cell proliferation was suggested as early as 1914. However, on the whole, these descriptions, along with those of pathologists looking at tissues in the first part of the twentieth century, described and thought of death as a passive process observed in sections of tissues by the morphological appearance of the cells. The descriptions given involve the condensation of chromatin and the generation of structures that were referred to as *pycnotic* nuclei. The fragmentation processes were also recorded and referred to as *karyorrhexis* and *karyolysis* for the degeneration of the nucleus and cytoplasm respectively.

Another good example, where apoptosis is regularly involved in the removal of perfectly healthy cells that are no longer needed, is in the human breast after lactation and during the monthly menstrual cycle. In the postlactation breast, functional cells are no longer required for milk production and the breast regresses in size to its prepregnancy condition. If pregnancy is reestablished, proliferation of the breast epithelium occurs and there is production of new, functional (milk-producing) cells. This is another clear example of cell death being initiated by bodily (systemic) signals (hormones) and the consequence is cell proliferation or cell suicide of unwanted cells. To a lesser extent, the phenomenon is seen in the breast at each successive menstrual cycle. As the breast prepares for pregnancy, proliferation occurs in the third week of the cycle, but in the absence of pregnancy the breast regresses and the cells produced by the limited amount of cell division in that cycle are lost by apoptosis at the end of each menstrual cycle.

The disturbance of the balance between cell production and cell loss that may well result in the growth of a tumour is not necessarily, and indeed is unlikely to be, an 'all or nothing' phenomenon. It is not the case that in a developing cancer there is no cell death or simply more cell proliferation, but that there is slightly less cell death or slightly more proliferation and, as a consequence, the proliferative compartment of the developing cancer can expand gradually with the passage of time. The balance between cell production and cell loss is tipped slightly in favour of cell production. Each new cancer cell (cancer stem cell) produced carries the same subtle genetic defect in the proliferation/cell death balance and so it and its progeny stand a slightly lower chance of dying (undergoing apoptosis) and a higher chance of proliferating. Consequently, although apoptosis would be expected to occur in tumours and indeed it is commonly observed, the implication is that in tumours the levels of cell death relative to the levels of cell proliferation are slightly lower than normal.

Mitotic death

A phrase that one occasionally encounters in the literature in relation to cell death is *mitotic death*. This is a term that is somewhat poorly defined and involves various processes and often means different things to different people. One interpretation is as follows: A cell may enter mitosis with certain levels of genetic damage, for example, chromosomal rearrangements that are incompatible with a complete and successful segregation of chromosomes at anaphase. As a consequence, the cell cannot complete the mitosis, aborts, and dies (via apoptosis) and this occurs within the mitotic phase. Another interpretation is a cell may enter mitosis with levels of damage to the DNA and chromosomes that *are* consistent with the segregation of chromosomes to the two poles of the mitotic spindle and the subsequent formation of two new daughter cells, but one or both of these cells then may carry damage that is recognised by the cell in G1 and the level of damage is such that cell death is induced. This type of cell death involves elements of damage detection and response and could, therefore, be regarded as involving aspects of cell programming and the death may morphologically resemble apoptosis. Alternatively, the cell may complete mitosis, but one of the two daughter cells may have lost part or all of some chromosomes because they were not effectively attached to the tubulin of the mitotic spindle because, for example, they lack the chromosomal centremere or attachment structure, in which case they tend to form small nuclei (micronuclei) in the cytoplasm. If one cell daughter lacks a chromosome(s), the other may contain an extra chromosome(s) as a micronucleus. In either case there is an imbalance, which, depending on the chromosomes involved and/or the amount of chromosome material and the genes carried thereon, may be incompatible with survival of the cell. Certain cancers and leukaemias are characterised by such chromosomal inbalances. This instability is induced at the time of mitosis and so this is regarded as one form of mitotic death (micronuclei can easily be confused in terms of recognition, with apoptotic fragments). Finally, certain agents may be toxic exclusively to cells when they are in the mitotic phase. The anaphase spindle poisons such as vincristine and vinblastine and colcemid and colchicine given at appropriate doses will kill cells in midmitosis. When this is done the cells die exhibiting a number of the morphological characteristics associated with apoptosis, but this is exclusively a death of cells from the mitotic phase. It is unclear why these forms of mitotic death should be separated from cell death induced in, or occurring at, any of the other phases of the life of a proliferating cell. There are a number of cytotoxic agents that will kill cells while they are replicating their DNA and one could refer to this as replication, or DNA synthesis or S-phase death, but on the whole this tends to confuse rather than clarify the situation.

Apoptosis versus necrosis

The distinction between apoptosis and necrosis is not always as clearcut as many would like to believe. In extreme circumstances the distinction is certainly clear: Necrosis tends to be induced by those sorts of processes that have a radical and immediate effects on cell metabolism and cellular machinery and the types of exposures that affect large numbers of cells simultaneously. The examples are respiratory poisons and physical disruption of cellular membranes caused by changes in the local osmotic environment, extreme pH, rises in temperature, and physical trauma. Metabolic poisons such as cyanide instantly prevent the cell from maintaining sufficient levels of its essential energy source, ATP. ATP is essential for the cell to maintain its salt/water balance, amongst other things. Under these circumstances, clusters of cells rapidly suffer changes; the cell membranes become leaky and ions can cross the cell membrane into the cell. Sodium ions and water are commonly accumulated and as a consequence the cells swell, calcium ions may leave the cell, and at later stages other cytoplasmic components, notably certain enzymes, can leak out of the cell and affect the surrounding tissue. Cellular metabolism stops and the DNA begins to be degraded in a completely nonspecific fashion. Internal membranes are also affected and, as a consequence, structures such as liposomes, which are intracytoplasmic bags or vesicles containing potentially dangerous enzymes, rupture and the enzymes are released into the cytoplasm and can also leak out of the cell into the tissue. Because of these leakage problems, the cells of the tissue recognise that severe damage has occurred and a cleanup operation is mounted that involves the infiltration of inflammatory blood cells into the tissue. As a consequence of the defects in the cell membrane, necrotic cells will absorb certain dyes and can, therefore, be stained. This is the basis of a technique known as 'vital staining', where healthy cells exclude the dye and necrotic cells take it up. There are a variety of 'vital stains' that could be used to distinguish between living and dead cells, based on the integrity of the cell membrane.

Unfortunately, the later stages of the digestion and degradation of apoptotic cells can morphologically resemble cell necrosis, making a distinction difficult on the basis of morphology, or appearance, alone. The situation is further complicated in that necrosis does appear to occur naturally under some circumstances. One situation where it is observed is during the process of spermatozoa production in the testes. During this process of sperm production, known as spermatogenesis, large numbers of cell divisions occur to produce large numbers of potential spermatozoa; however, many of these cells die, some by apoptosis but some by necrosis, at various stages in the spermatogenic sequence.

DNA degradation

In contrast to necrosis, the cell membranes remain intact during apoptosis (there is no leakage of enzymes out of the cell) and metabolic activity continues. The DNA is degraded in an organised fashion by a calcium- and magnesium-dependent DNA cleaving enzyme called an *endonuclease*. The double helix of the DNA is supercoiled and folded in a complex fashion into loops and rosettes (see figure 7). As mentioned previously, at regular intervals the DNA is wound around specialised, chromosome-associated proteins called histones to form small structures called nucleosomes (figure 8). These nucleosomes are regularly spaced along the DNA strand about 200 DNA bases apart. Because the DNA is a double helix with two DNA strands this is usually referred to as 200 base pairs (or pairs of DNA

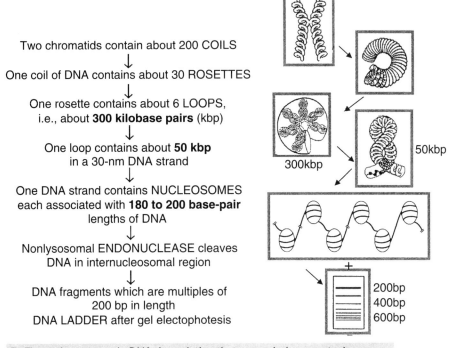

Two chromatids contain about 200 COILS
↓
One coil of DNA contains about 30 ROSETTES
↓
One rosette contains about 6 LOOPS,
i.e., about **300 kilobase pairs** (kbp)
↓
One loop contains about **50 kbp**
in a 30-nm DNA strand
↓
One DNA strand contains NUCLEOSOMES
each associated with **180 to 200 base-pair**
lengths of DNA
↓
Nonlysosomal ENDONUCLEASE cleaves
DNA in internucleosomal region
↓
DNA fragments which are multiples of
200 bp in length
DNA LADDER after gel electophotesis

7. The various stages in DNA degradation that occur during apoptosis.

bases). The endonuclease, which is activated from a preformed inactive state, cuts the DNA molecule in the region between nucleosomes and consequently DNA fragments that are essentially multiples of 200-base-pair lengths are generated. The DNA can be extracted from cells undergoing apoptosis and the DNA fragments isolated on the basis of their size. When this is done, the DNA extracted from apoptotic cells shows a characteristic, banding pattern with a very regular spacing that is consistent with a large number of fragments of DNA of common sizes ranging from simple to multiple complexes of 200 base pairs in length. This characteristic pattern of bands is referred to as a *DNA ladder*. An example of this is shown in figure 9. If the same technique is used on cells simultaneously undergoing necrosis, DNA fragments of an infinite range of sizes are generated and hence no discrete bands or ladders can be visualised – just a smear.

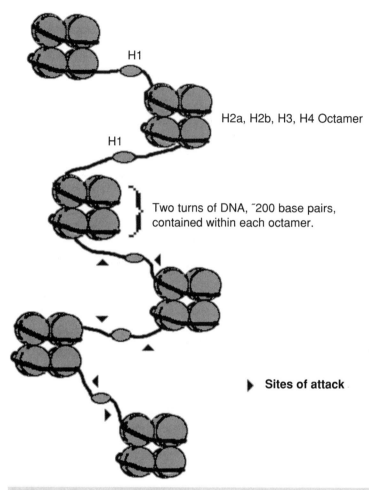

8. Diagram showing a DNA strand coiled around specialised proteins (histones, H1–H4) to form structures called nucleosomes. Arrowheads indicates the sites of attack by the apoptotic endonucleases, which result in the formation of DNA fragments of discrete lengths.

Until recently, the presence of a DNA ladder was thought to be a definitive biochemical marker of apoptosis and this is still true in many situations. However, it has become clear that in some apoptotic cells endonuclease-mediated degradation of the DNA down to 200-base-pair fragments does not occur, although degradation of the DNA into 300,000- and 50,000-base-pair size fragments is apparent. It is believed that these fragments are generated by the enzyme *topoisomerase II*. These significantly larger DNA fragments can also be

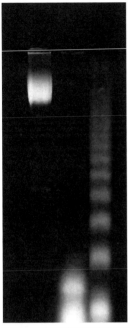

9. Characteristic 'laddering' pattern of DNA isolated from apoptotic cells, which has been subject to separation using agarose gel electrophoresis. The size of the fragments determines their speed of migration through the gel matrix, with the smallest moving the most rapidly and consequently appearing further down the gel. The discrete sizes arise from the random cleavage at the specific internucleosomal sites (see figure 7). In the first lane (column) of the gel, a set of marker fragments of known molecular size has been run (courtesy of Dr. K. Tonaka, Osaka).

easily recognised using appropriate DNA extraction and analysis techniques. The 50 kilobase fragments correspond to loops of DNA and the 300 kilobase fragments correspond to the rosette structures made up of six loops, as illustrated in figure 7. In all cells undergoing apoptosis then, some or all of these very specific sizes of DNA fragments are generated and this is a major distinguishing feature between apoptosis and necrosis.

An early stage in the process of necrosis may involve the denaturation (breakdown) of cellular proteins. Denaturation of the histones would breakdown the nucleosomal structure and allow the breakage of the DNA at any point along its length. There is no activation of a specific endonuclease in cells undergoing nercrosis; however, many enzymes will cut DNA and each has its own specific cutting point.

Consequently, fragments of random size will be obtained, and not just those of nucleosomal unit lengths.

Why is it that during the process of cell death (necrosis or apoptosis) the DNA is systematically broken up? One possibility here is that this is part of nature's efficient recycling process. Much effort has been undertaken by the cell in making DNA, its most important molecule. If cells are deleted or die these basic building blocks of DNA could be efficiently reutilised and recycled. There is considerable evidence that the individual bases, and even fragments of DNA, from cells at the end of their differentiated life span are released from the dying cell and reutilised and reincorporated into the DNA of proliferative cells that are replicating their DNA. This can be demonstrated in the intestine when epithelial cells with DNA that has been prelabelled reach the end of their life span and are shed into the centre of the intestinal tube. The labelled bases can be seen to be reincorporated after an appropriate time into the proliferating cells in the crypts. This can also be seen in the epidermis in the skin. Here, in the last stages of the differentiated life span, before the functional skin surface cells are shed, the cell carefully dismantles its nucleus and in the course of doing so generates DNA bases that can be reincorporated into the DNA of the proliferative cells deeper in the epidermis. This dismantling of the nucleus in the upper layers of the skin is thought by some to have some similarities with the process of apoptosis. Hence, the process of dismantling the DNA using endonuclease enzymes may be important for recycling or reutilising the constituent bases or even small fragments of DNA in a variety of tissues.

How do we recognise apoptosis?

The ease with which apoptosis is recognised in various situations depends firstly on the frequency of apoptotic cell death amongst the cell population of interest and secondly on the techniques used to recognise cells that are in the process of apoptosis. The techniques used tend to vary depending on whether cell cultures, whole organisms, embryos, or sections of tissue are being considered. The techniques that are available for identification of apoptosis broadly fall into four categories as follows: (1) methods for assessment of DNA integrity, (2) methods to assess the activation of apoptosis-specific proteolysis, (3) methods for determining changes in the levels of key proteins that regulate the apoptotic process, (4) methods to assess membrane changes, and (5) studies of morphological change. The techniques employed in assessing these parameters may range from simple light microscopy to classical biochemistry to the use of sophisticated and very expensive machines, such as flow cytometers.

Assessment of DNA fragmentation

As mentioned previously, the cell's DNA is cleaved during apoptosis into fragments of specific sizes. The DNA can be extracted from the cells by lysing them in a hypotonic salt solution. The cellular protein and RNA are then digested, prior to separating the fragments, on the basis of size, by placing the DNA in a gel matrix and applying an electrical potential difference. This technique is called *agarose gel electrophoresis*. The DNA is negatively charged and will migrate towards the anode. Commonly the gels are between 10 and 25 cm in length and the potential applied is 40–100 V; it will take three to four hours to separate out the internucleosomal fragments. The gel is then stained with a DNA-binding dye, ethidium bromide, which fluoresces under ultraviolet illumination, allowing the DNA fragments to be seen. More complex separation protocols, in which both the size and orientation of the potential applied are varied, can be used to visualise the large 300- and 50-kilobase DNA fragments that also occur during apoptosis. This technique is called *pulsed-field gel electrophoresis*. This technique may be applied to cells grown in culture or cells isolated from whole tissue and even homogenised tissues, if there is sufficient apoptosis within the tissue as a whole.

Another method for assessing DNA fragmentation is to measure the *fractional DNA content* of the cells. For this it is necessary to prepare a single-cell suspension. If it is to be used on tissue samples, this may involve various disaggregation and enzymatic digestion steps to isolate the cells. The cells must then be fixed, permeabilised, and labelled with a DNA-binding fluorescent probe, commonly propidium iodide (which fluoresces red). Like the preparation for agarose gel electrophoresis, the RNA must be degraded, as this will also be tagged by the fluorescent probe. The amount of probe that is bound within any given cell, and hence the fluorescence of the cell, will be proportional to the amount of DNA in that cell. If cells with the normal DNA content of $2n$ (forty-six) chromosomes are defined as having a fluorescence of 1, a cell in the G2/M phase of the cell cycle, which has completed replicating its DNA and has $4n$ chromosomes, will have a relative fluorescence of 2. Cells in S phase, which are in the process of replicating their DNA, will have a fluorescence value between 1 and 2. Cells in which DNA degradation has taken place (i.e., apoptotic cells) will have a relative fluorescence value of less than 1. The fluorescent labelling of the cells can be assessed using a flow cytometer (see figure 10).

The flow cytometer is a very sophisticated machine. It takes up the cell suspension and generates a single droplet suspension (the droplets being of a size equivalent to a single cell), which passes down a narrow chamber, through an incident beam of laser light. Different lasers can be used as sources of specific wavelengths of monochromatic light to excite a range of fluorescent dyes. The light will be deflected or reflected by the contents of the drop depending on its light-scattering properties, which are determined by the size and density of the cell. Voltages are then applied to focus the light emission from the dyes onto photon detectors. In this way, the relative fluorescence of each individual cell can be assayed. The user can programme the number of events that are counted (commonly 10,000) and even the size of the objects that are analysed. Dedicated computer software can then be used to collate and analyse the data to produce frequency distributions relating to the size and the fluorescence of the cells. A typical frequency distribution of the DNA content of proliferating cells in culture will show two peaks (representing cells in G1 and cells in G2/M) separated by a plateau or variable region (cells in S); apoptotic cells will appear as a peak or hump to the left of the first peak (G1), which is often referred to as the

HOW TO DIE?

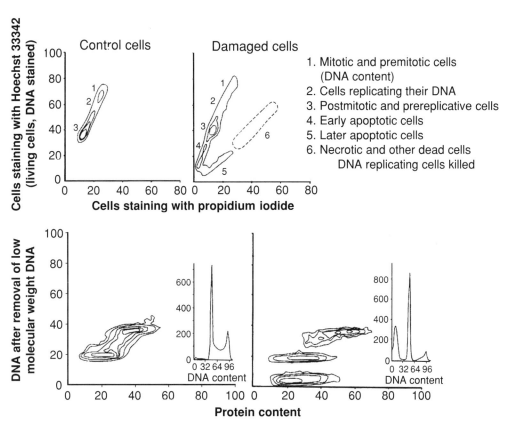

10. Analysis of data from flow-cytometric analysis of normal and apoptotic cells. Top two panels show unfixed cells that have been stained with propidium iodide and subsequently permeabilised and stained with Hoechst 33342 dye. The propidium iodide is excluded by live cells and only taken up by cells as they lose their plasma membrane integrity (i.e., late apoptotic and necrotic cells). The amount of Hoechst 33342 dye taken up is proportional to the DNA content of the cell, so the cells replicating their DNA will stain more than those that are not. The bottom two panels show simultaneous analysis of protein and DNA content. Note that apoptotic cells have reduced DNA content, due to degradation (modified and redrawn from Darzynkiewicz *et al.*, analysis of cell death by flow cytometry. In: *Cell Growth and Apoptosis: A Practical Approach*, G. P. Studzinski (ed.). Oxford University Press, Oxford, 1995).

subdiploid peak. Although this technique permits automated analysis of thousands of cells, it will give no information on the location of apoptotic events when used to analyse cells isolated from whole tissues.

The other commonly used technique for assessing DNA degradation in apoptotic cells is to use an enzyme to 'tag' all the cuts in the

DNA that have been made by the endonuclease. There are two main, but subtly different methods that are currently employed. They are *in situ* end labelling (*ISEL*) and terminal deoxynucleotide transferase (TdT) mediated dUTP-biotin nick end labelling (*TUNEL*). The techniques differ with respect to the enzyme employed to tag the DNA breaks. TUNEL employs the TdT enzyme, whereas ISEL uses DNA polymerase. Both serve to incorporate a single, extra DNA base onto the broken ends of the DNA strands where the cuts have been made by the endonuclease. DNA polymerase can label ends of DNA only where both strands of the DNA have been cut right through; however, TdT can also label the ends of DNA where only a single DNA strand has been cut. The TUNEL technique is consequently more sensitive than ISEL. In order for the enzyme to gain access to the nucleus and to the DNA strand breaks, pretreatment of cells and tissues, in addition to any other fixation techniques, is required. This treatment may be enzymatic (usually the proteinase protease K is used) or detergent-based.

The labelled nucleotide may be visualised by different methods according to its tag. A biotin tag will bind the protein avidin in a multivalent manner: up to four avidin molecules may bind to one biotin molecule – this allows amplification of the signal. The avidin, in turn, is bound either to an enzyme like horseradish peroxidase (HRP) or to a fluorescent molecule, such as fluoresceine or Texas Red. Fluorescent molecules may be visualised directly using a fluorescence microscope. With regard to enzyme detection methods, a substrate is added to the cells; this is converted to a coloured product, which is subsequently deposited around the area of the DNA strand breaks. 3'3'-Diaminobenzidine (DAB) is a commonly used substrate for HRP and has a brown product that can be visualised using light microscopy (figure 11). Digoxigenin- (an insect protein) labelled DNA bases are commonly used in ISEL. Digoxigenin is detected using a specific, antidigoxigenin antibody, which does not react with any mammalian proteins. Like TUNEL, the antibody may be tagged with a fluorescent molecule or with an enzyme and the detection then proceeds via the method outlined above.

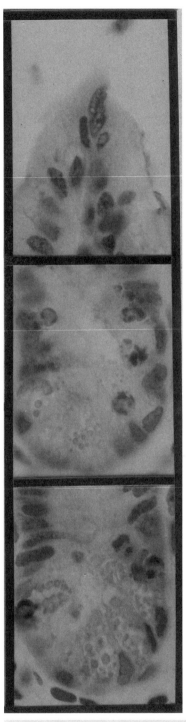

11. TUNEL assay for apoptosis in intestinal epithelium. Upper panel shows the functional cells at the tip of the villus (see Chapter 7). The lower two panels show apoptosis in the proliferative compartment, the crypts. Cells may be observed with characteristic apoptotic morphology (arrowheads). Most of these cells also show brown stain (TUNEL). Cells in division (mitosis) are shown by arrows.

This technique is used on cells grown in culture and on thin (5 μm or less) sections of tissues. In addition to microscopy, fluorescently labelled cells in suspension can be analysed using a flow cytometer. Many kits to perform ISEL and TUNEL are now available commercially, offering a wide range of detection options and some claiming to have been optimised for specific tissues.

The difficulties with these techniques are that very different staining patterns seem to be obtainable with different cell preparations and tissues. This may be because these different biological samples have a differing sensitivity to the enzymatic digestion step involving proteinase K. Some cell and tissue preparations do not tolerate the proteolytic digestion at all and detergent may be used instead. The TdT enzyme activity has a dependence on cations and the type of cation used (either cobalt, manganese or magnesium) may be varied according to tissue type in order to give optimal activity. What this tends to mean in practice is that the techniques have to be extremely carefully standardised and validated for almost each individual situation.

The second problem is that the assessment and recording of the number of positively stained cells by light or fluorescence microscopy is subjective. There will often be some background staining of cells using either the HRP labelling or fluorescence labelling techniques. How does one set a threshold for describing a cell as labelled (i.e., what is significantly brown or fluorescent red, for example)? Different observers commonly set individual thresholds that may differ one from another. Digital cameras and sophisticated image capture and analysis software for fluorescence microscopy now allow the observer to set defined thresholds for observation. Also, analysis of labelling (by fluorescence) of isolated cells using a flow cytometer allows the user to programme thresholds for detection and allow the user to define what the background level of staining really is.

The final problem is that it remains unclear to what extent these end-labelling techniques permit a discrimination amongst apoptosis, necrosis, DNA damage induced by cytotoxic agents (DNA damage may be repaired and not necessarily result in execution of the apoptotic programme), and DNA deterioration due to the age of the cell.

DNA strand breaks are also reported to be present in cells with high levels of transcriptional activity. So, these techniques may not be very specific markers for apoptosis under certain circumstances.

If one wishes to address the question of how specific the particular approach is for apoptosis a reference standard needs to be established. At present, this is most likely to be the morphological appearance of the cell. When this standard is used it is clear that these techniques identify a variable proportion of normal-looking cells that stain positively for strand breaks. These could be false nonapoptotic positive cells. Alternatively, they may represent cells early in the apoptotic process (i.e., before any morphological change has occurred). It is also clear that a variable proportion of cells exhibiting the classical morphological features of apoptosis may be unstained using the strand-break techniques. These cells are false negatives or apoptotic cells that do not stain for broken DNA. They may remain unstained because the DNA has become so condensed that there are no free ends to label. Very few people using these end-labelling techniques actually validate the technique and determine the proportion of false negatives and false positives. However, the advantage of these techniques is that they can, in principle, be combined with various other approaches, and when this is done they can be valuable techniques for identifying apoptotic cells.

Assessment of protease activity in apoptosis

Apoptosis is associated with the activation of a specific family of protease enzymes called *caspases*. These are discussed in greater detail in Chapter 5. Here an enzyme is defined as an intracellular or secreted protein that binds to other proteins or molecules in a highly specific manner and cause or accelerate a change in the bound molecule (i.e., they act as catalysts for chemical change). Caspases are calcium-dependent proteases that cleave after aspartic acid residues. The name is an acronym for cysteinyl aspartase. Several approaches are used to determine the activation of these apoptotic proteases: these include biochemical assays and immunologically-based assays.

Each caspase has a unique amino acid sequence that it preferentially cleaves and polypeptide substrates have been designed that are specifically cleaved by certain caspases. The substrates are linked to a tag molecule that, when cleaved by the action of the caspase, gives a coloured or fluorescent signal that may be measured using a spectrometer. The intensity of the colour change, or fluorescence signal, is related to the activity of the specific caspase. This technique is most suitable for cells grown in culture. It is commonly performed on extracts from lysed cells, although cell-permeant dyes have been developed that allow activity to be assessed in living cells.

Antibodies are also available that can recognise one caspase, casapse-3, specifically in its active form. Caspases usually exist in an inactive form in cells, but upon induction of apoptosis, they are cleaved either by autocatalysis or by other caspases. The shortened molecules then dimerise (pair) to produce the active form of the enzyme. The antiactive caspase-3 antibody may be applied to cells that have been fixed and permeabilised or to fixed tissue sections. A second antibody is then applied that specifically binds to the type of immunoglobulin from the particular species in which the first antibody has been raised: antibodies are commonly of rabbit or mouse origin. The secondary antibody may be tagged with biotin, with an enzyme (e.g., horseradish peroxidase), or with a fluorescent probe and detection may be carried out by the methods described previously for the end-labelling techniques. These techniques are called *immunocytochemistry* when the detection technique results in a visible stain assessed by light microscopy and *immunofluorescence* when fluorescent probes and fluorescence microscopy are used.

The other approaches to assess caspase activation are also immunologically based. As mentioned, caspase molecules are cleaved during their activation. If one prepares a cell lysate, the proteins within the cell can be separated on the basis of the their size, in much the same way as DNA fragments, by loading them onto one end of a gel matrix and applying an electric current. The gel used is made of polymerised acrylamide and bis-acrylamide; the separation technique is termed *polyacrylamide gel electrophoresis* (PAGE). Often, it is essential to cool the gels as the proteins are being separated, as much heat is generated. After the proteins are separated, they are transferred onto a microporous membrane (either cellulose-based or synthetic), by applying an electrical potential difference across the juxtaposed membrane and gel, a process called *Western blotting*. Antibodies can then be applied to the membrane that recognise specific caspases (or other proteins). As with the detection of active caspases, a secondary antibody is applied: here it is common to use an antibody tagged with horseradish peroxidase. The membrane is then incubated with a substrate solution such as DAB to given a coloured product on the membrane or, more commonly, with a substrate that gives a chemiluminescent (photon-emitting) product that can be detected by placing the membrane against light-sensitive film in the dark for a short time, after which it is developed. After this, one is left with membrane, or photographic image of the membrane, with discrete bands that correspond to the proteins recognised by the primary antibody. The size of the bands can be confirmed by running samples containing a cocktail of proteins of known molecular size alongside the experimental samples. In nonapoptotic samples caspases will primarily exist in their precursor form and so just one band will be seen; however, in samples from apoptotic cells two additional smaller bands may be seen corresponding to the fragments of the cleaved precursor. As mentioned previously, the fragments are recombined and dimerise with other cleaved caspase molecules to form the active enzyme; therefore, conditions are used in the preparation of the samples and in the electrophoresis that breakdown multisubunit proteins into their component parts. In this way, the presence of the smaller caspase fragments can be used as a reliable indicator of caspase activation.

This is quite a widely used technique and can be applied to cells in culture and whole tissues, although it is a little uninformative when used on whole tissues as it will not indicate which cell types within that tissue are undergoing apoptosis. Also, if the apoptotic cell population in a tissue is not a significant proportion of the tissue mass, the levels of active caspase fragments may be difficult to detect as there will be a great excess of nonapoptotic cells in the tissue contributing to the protein in the sample. It is better, therefore, to isolate specific cell populations from a tissue when using this technique. Another caveat is that caspase activation also occurs in some nonapoptoptic contexts in certain cell types.

The same techniques of PAGE and Western blotting may be applied also to examining the cleavage of specific caspase substrates. There are many proteins that are specifically cleaved by caspases during apoptosis; many are proteins found in the nucleus of the cell and are involved in maintaining the DNA. One such example and one of the first to be described experimentally is *poly-ADP ribose polymerase* (PARP) and the appearance of cleavage fragments of this enzyme have been shown to closely follow the time course of apoptotic events.

Changes in apoptosis-regulatory proteins

There are many proteins that regulate apoptosis and their levels may go up or down, or their state of activation or cellular localisation may change during apoptosis. These include members of the Bcl-2 family of proteins (see table 3), named after the first member to be discovered (Bcl is an acronym for B-cell lymphoma), and the protein p53, which orchestrates the response of the cell to DNA damage. Bcl-2 suppresses apoptosis and promotes survival and its expression is often decreased during apoptosis in many cell systems. Other members of the Bcl-2 protein family promote apoptosis and increase following the induction of apoptosis, these include proteins called Bax and Bak. p53 levels also increase following damage to a cell's DNA and, according to how much it is increased, may either initiate arrest of cell cycle progression or induce apoptosis, through regulating the expression of specific sets of genes in the cell.

PAGE and Western blotting can be used to assess changes in the levels of proteins in cells. At best this gives a semiquantitative estimate of protein levels and also only gives an 'average' of the level of expression in the cell population as a whole, as the technique is performed using protein extracted from cell and tissue lysates involving millions of individual cells. More quantitative assessment can be performed by using fixed and permeabilised, isolated cell populations and flow cytometry to assess expression of specific proteins that have

TABLE 3
Bcl-2 Family proteins

Antiapoptotic	
Mammalian	Bcl-2, Bcl-x_L, Bcl-w, Mcl-1, BAG1, Bfl-1, Boo,
Viral	E1B-19kD, BHRF-1, KSBCL-2
C. elegans	CED-9
Proapoptotic	
Mammalian	Bax, Bak, Bik, Bok, Bad, Harakiri, Bcl-x_s
C. elegans	Egl-1
Drosophila	dBorg-1

been labelled using immunofluorescence techniques, as previously described.

Immunostaining techniques can also be used to monitor changes in not only the expression of specific proteins, but also the change in their cellular distribution. PAGE and Western blotting may be able to demonstrate an increase in, for example, the level of p53 expression following DNA damage, but using microscopy and immunostaining techniques will show that this increase occurs specifically in the cells' nuclei. The former techniques can be made more specific by a process called subcellular fractionation, which relies on high-speed centrifugation to separate out the different cellular compartments (e.g., the cytosol, the nucleus, the mitochondria, and the membrane fraction). These separate bits of the cell can then be used to prepare protein extracts and used for PAGE and Western blotting. Subcellular fractionation can be used, therefore, to show the movement of a protein from one cellular compartment to another. An example is cytochrome *c*, a respiratory chain protein that resides in the mitochondria that is released into the cytosol during apoptosis. Cytochrome *c* movement out of the mitochondria is dependent on pores opening in the mitochondrial membrane. These pores open due to the action of proapoptotic Bcl-2 family proteins, such as the protein Bax, which relocate from the cytosol to the mitochondrial membrane. The movement of Bax and cytochrome *c* can be visualised at very high resolution using immunofluorescence to detect the proteins in fixed cells or tissue sections and then visualising them using a sophisticated microscope, called a fluorescence confocal microscope. This scans through the entire thickness of a cell or tissue section with a laser beam and records the emission from the fluorescent probes applied to that cell/section. It can then use computer software to generate a 3D image of the cell. It is also much more sensitive than a conventional fluorescence microscopy. Recently, it has become possible to observe the movement of such proteins in real time in cells in culture that are actively undergoing apoptosis by using molecular and gene transfer techniques to make cells that express proteins tagged with another fluorescent protein, appropriately called green fluorescent protein (GFP). These GFP-tagged proteins are beginning to allow us to study many new and different aspects of biology.

Membrane changes

Vital stains

Vital dyes do not enter live cells with intact plasma membranes. They include dyes that are visible using light microscopy and also dyes that are fluorescent. These dyes may be used on their own to assess cellular change, but are most useful in combination with flow cytometric analysis of other cellular parameters. Dying cells, specifically necrotic cells and those in the very late stages of apoptosis, have damaged cell membranes and can be stained with dyes such as *trypan blue* or *propidium iodide* (fluorescent red). An alternative approach in the use of dyes is to use one that is taken up by living cells. One example is *fluorescein diacetate*, which is metabolised by the living cell to generate a dye called fluoroscein that can be seen in the cytoplasm, which fluoresces green. Using fluoroscein diacetate and propidium iodide together will result in living cells appearing green and, under the appropriate illumination, dead cells appearing red.

Changes also occur in the membranes of intracellular organelles during cell death, most notably the mitochondria as just described. *Rhodamine 123* and the more recently developed dye *JC-1* are fluorescent dyes (positively charged) that accumulate within the mitochondria (which have a strong negative charge) of living cells, where they aggregate and fluoresce green. As mentioned, during apoptosis, mitochondrial proteins such as cytochrome *c* are released into the cytosol, due to the opening of pores by proteins such as Bax. Likewise the dyes will also leak out into the cytosol, where they disaggregate and fluoresce red/orange. Again they can be used in conjunction with propidium iodide to generate green living cells and red dead cells.

The lysosome is a cytoplasmic organelle containing a variety of enzymes. *Acridine orange* is selectively (although not exclusively) taken up by the lysosomes in living cells and will fluoresce red. The DNA specific stain known as *Hoechst 33342* can cross the cell membrane and stain DNA in living cells, again this can be used in conjunction with other stains.

As mentioned, cells stained with these dyes can be analysed with flow cytometry and their staining properties assessed in relation to their physical size. As mentioned previously, in a flow cytometer the cells pass as a single cell suspension through an incident beam of laser light. Apart from any fluorescent probes emitting light of a specific wavelength, the incident light is reflected and scattered by the cells. The amount of light reflection and scatter is dependent on the physical properties of the cell. Large cells tend to reflect more light, whereas dense object scatter more light. However, cell debris (broken fragments of cells) and necrotic cells also tend to have reduced light-scattering properties. The use of vital dyes can allow the researcher to discriminate these objects. Figure 10 demonstrates the combined use of two of the dyes just mentioned and the pattern obtained for a frequency scatter plot from which apoptotic cells can be identified. Other examples using a combination of techniques with the flow cytometer are also shown in figure 10.

Changes in organisation of plasma membrane components occur during apoptosis. The plasma membrane consists of several different phospholipids. One, phosphatidylserine (PS), is kept exclusively within the inner leaflet of the membrane bilayer by an enzyme, which makes PS flip back to the inside layer whenever it appears in the outside layer. In cells undergoing apoptosis, this mechanism is disabled and PS appears in the outer leaflet of the membrane, where it acts as a signal to other cells to phagocytose the apoptotic cell. Within the body there is an anticoagulatnt molecule, *annexin-V*, which shows strong affinity for binding to negatively charged phospholipids like PS. This property of annexin-V has been exploited to devise another method for detecting apoptotic cells. Annexin-V can be conjugated with a fluorescent tag (e.g., fluorosceine) and incubated with cells stimulated to undergo apoptosis. The labelled annexin-V will bind preferentially to those cells that are in the process of undergoing apoptosis and, as a consequence, have PS in the outer leaflet of their plasma membrane. The annexin-V that has not bound to any cells can then be washed away and flow cytometry or fluorescence microscopy can then be used to determine the relative number of cells that are fluorescent (i.e., have bound annexin-V and are apoptotic).

Morphology

Finally, we come to what is regarded as the most traditional approach to studying apoptosis and for many researchers the most definitive, characterising the morphological changes the cell is going through (i.e., what does it look like?). This technique relies on the use of the microscope. By far the greatest sensitivity for detection of apoptosis is obtained in the electron microscope and impressive photomicrographs illustrating the morphological changes can be readily obtained. However, the electron microscope is not a practical piece of equipment for determining quantitative parameters (i.e., how many cells are dying in tissues), because only very small fields and few cells can be analysed at any one time. The changes that have already been described in earlier sections can be relatively easily recognised in appropriately stained histological preparations (sections) or samples involving whole cells spread on a microscope slide and then analysed with a light microscope (see figure 12). A wide variety of stains can be used, the most common of which is probably haematoxylin and eosin; this can be viewed using conventional light microscopy. These may require the use of ultraviolet light to obtain specific wavelength emissions from the dyes or straightforward transmitted light observations. One of the limitations determining the number of cell death events that occur in such preparations is the ability to distinguish between apoptosis and necrosis at the magnifications used in the light

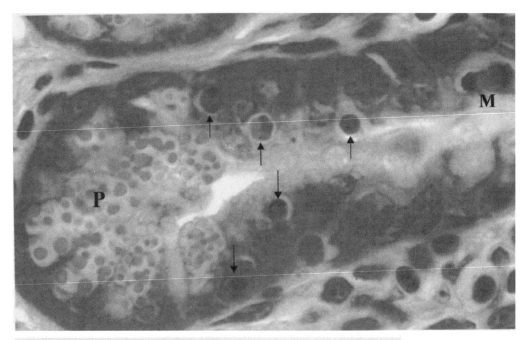

12. Haematoxylin-and-eosin-stained section through a small intestinal crypt showing many apoptotic fragments (arrows), a mitotic cell (M), and functional differentiated cells, Paneth cells (P).

microscope and the thresholds for detection of the event (i.e., how easy it is to recognise the early initial phases and the ability to discriminate small fragments). The latter is dependent somewhat on the magnification used for the observation. However, this type of approach is still the best reference standard for most of the other practical approaches. There are, however, a number of difficulties concerning the interpretation of the measurements that are made and these are discussed in Chapter 9.

Cell death in cell cultures

One situation where cells die through an apoptotic sequence of events is in cell cultures. Epithelial and endothelial cells must form an attachment in a culture dish to ensure cell survival. They cannot generally grow in a freely suspended state in culture medium (suspension cultures). When they are growing attached to the bottom of a culture dish and enzymes are used to break their attachment they die. Cells *in vivo* and *in vitro* are also continuously 'talking' to each other using chemical signals as messages.

This requirement for close physical contact with other cells may be an important feature of such cells *in vivo*. It may help ensure a strong and effective barrier between the inside of the body and the hostile outside environment. It may also reduce the likelihood of cells detaching and migrating or being carried to other regions of the body and then growing in an abnormal location. This is essentially what metastasising cancer cells do. It may also be important for the exchange of survival signals that keep cells from dying. Fibroblasts in contrast are the major cell type in the supporting connective tissue and these can grow without attachment. When cells in culture, attached to the culture dish, are exposed to agents that kill cells they tend to round up and detach themselves. These dying cells then float in the culture medium. Their numbers can be determined to provide an index of the cytotoxic efficiency. Various studies have been performed to demonstrate that these free-floating cells exhibit some, or many, of the properties of apoptotic cells. However, not all the attached cells may be fully functional living cells, for example, some may have a reduced potential for cell division being either incapable of further division or having a restricted division potential. Such "undead" cells are sometimes regarded as postmitotic or prematurely differentiated if they have completely lost their division potential. Those with a reduced or limited division potential will form abortive small colonies when grown under conditions where the true survivors can produce colonies exceeding fifty cells within a prescribed time scale. Thus under such conditions following a cytotoxic exposure (e.g., radiation), there will be surviving reproductively competent cells and reproductively impaired or sterilised cells attached to the culture dish and free-floating dead (apoptotic) cells. The term *anoikis* has been coined to describe this loss of attachment type of cell death. It is derived from the Greek word meaning homelessness and is pronounced *anoeekis*. An *in vivo* situation where a similar cell detachment process or anoikis is seen is at the tip of the villus in the small intestine where cells are detached or extruded into the lumen of the intestine, at the end of their functional life span. These cells have probably initiated at least part of the apoptosis programme because some of the cell death-associated genes appear to be activated prior to extrusion. There are various cell membrane-associated receptors called intergrins and attachment proteins called cadherins that are involved in the regulation of this anchorage-dependent cell survival or death.

Further reading for Chapters 1 and 2

Arends, M. J., Morris, R. G., and Wyllie, A. H. Apoptosis: The role of the endonuclease. *Am. J. Pathol.* **136**: 593–608, 1990.

Bowen, I. D., and Bowen, S. M. Programmed Cell Death in Tumours and Tissues. Chapman & Hall, London, p. 268, 1990.

Bowen, I. D., and Lockshin, R. A. Cell Death in Biology and Pathology. Chapman & Hall, London, p. 493, 1981.

Dexter, T. M., Raff, M. C., and Wyllie, A. H. The Role of Apoptosis in Development, Time, Homeostasis and Malignancy. The Royal Society. Chapman & Hall, London, p. 101, 1995.

Harmon, B. V., and Allan, D. J. Apoptosis: A 20th century scientific revolution. In: Apoptosis in Normal Development in Cancer *(edited by M. Sluyser). Taylor & Francis London*, pp. 1–19, 1996.

Holbrook, N. J., Martin, G. R., and Lockshin, R. A. Cellular Ageing and Cell Death. *Wiley–Liss, New York*, p. 319, 1996.

Kerr, J. F. R., Wyllie, A. H., and Currie, A. R. Apoptosis: A basic biological phenomenon with wide ranging implications in tissue kinetics. *Br. J. Cancer* **26**: 239–257, 1972.

Merritt, A. J., Allen, T., Potten, C. S., and Hickman, J. A. Apoptosis in small intestinal epithelium from p53-null mice: evidence for a delayed, p-53 independent G2/M-associated cell death after irradiation. *Oncogene* **14**: 2759–2766, 1997.

Potten, C. S. The significance of spontaneous and induced apoptosis in the gastrointestinal intestinal tract of mice. *Cancer Metastases Rev.* **11**: 179–195, 1992.

Potten C. S. What is an apoptotic index measuring? A commentary. *Br. J. Cancer* **74**: 1743–1748, 1996.

Potten, C. S. Perspectives on Mammalian Cell Death. *Oxford Science Publications, Oxford*, p. 363, 1997.

Raff, M. C. Social controls on cell survival and cell death. *Nature* **356**: 397–400, 1992.

Sluyser, M. Apoptosis in Normal Development and Cancer. *Taylor & Francis, London*, p. 304, 1996.

Studzinski, G. P. (ed.). Cell Growth and Apoptosis: A Practical Approach. *IRL Press/Oxford University Press, Oxford*, p. 269, 1995.

Thomas, N. S. B. (ed.). Apoptosis and Cell Cycle Control in Cancer. *Bios. Scientific Publishers, Oxford*, p. 238, 1996.

Tomei, L. D., and Cope, F. O. (eds.). Apoptosis, the Molecular Biology of Cell Death. *Cold Spring Harbor Press, New York*, p. 321, 1991.

Wyllie, A. H. Apoptosis and the regulation of cell numbers in normal and neoplastic tissues. *Cancer Metastases Rev.* **11**: 95–103, 1992.

Wyllie, A. H., Kerr, J. F. R., and Currie, A. R. Cell death, the significance of apoptosis. *Int. Rev. Cytotol.* **68**: 251–306, 1980.

3

What to wear and who clears up the rubbish?

When a cell dies by apoptosis, the cell corpse and associated debris (i.e., apoptotic bodies) must be cleared up (figure 13). In much the way that throwing a dead body into the street would be offensive (it would smell and attract rats and be a health hazard), leaving apoptotic cells 'lying around' can cause problems for an organism. Failure to clear apoptotic corpses can result in inflammatory responses and the production of autoantibodies (antibodies that recognise 'self') in response to antigens derived from disintegrating apoptotic cells and any hypothetical cell death-related toxins. Most autoantibodies are against nuclear proteins that are specifically cleaved by the action of caspases during apoptosis.

13.

So, how are apoptotic cells cleared up? In the nematode worm *C. elegans*, two sets of three genes have been found that control two separate mechanisms for the phagocytosis of apoptotic cells. These sets of genes are *ced-1/ced-6/ced-7* and *ced-2/ced-5/ced-10*. Worms that possess a mutation of any one of these genes show defective engulfment (removal) of apoptotic cells, with the consequence that the 'corpses' of the dead cells remain within the body of the animal for an extended period. These cell corpses can be visualised using light microscopy techniques, appearing as obvious discs within the body of the worm, which is otherwise quite healthy. The apoptotic cells will eventually disappear, indicative of the functional redundancy that exists in the phagocytic clearance pathways.

In mammalian systems, some of the functional detail concerning the phagocytosis of apoptotic cells has been elucidated. It appears that apoptotic cells display specific 'eat me' signals on their surface and, in contrast, live cells are able to actively repel the attentions of phagocytic scavenger cells, such as macrophages. In fact, there are several systems involved in the recognition and engulfment of apoptotic cells. Currently, one of the best characterised is the display of the negatively charged phospholipid *phosphatidylserine* on the surface of apoptotic cells. In live cells, phosphatidylserine is usually confined to the innner leaflet of the plasma membrane lipid bilayer. It is kept there via the action of an enzyme that actively promotes the flipping back of the phosphatidylserine from the outer to the inner leaflet of the bilayer. This enzyme is inactivated during apoptosis and, hence, the phosphatidylserine appears on the outside of the cell.

Macrophages possess a receptor protein for phosphatidylserine on their surface. The recognition of the phosphadtidylserine molecules by this receptor initiates the phagocytosis of the apoptotic cells. Gene transfer studies have confirmed that this phosphatidylserine receptor can confer the ability to phagocytose (engulf) apoptotic cells on cell types that do not usual possess this ability, such as T and B lymphocytes.

A macrophage will break down the cells or microorganisms that it ingests and display small pieces of digested protein (eight to twelve amino acids in length) on its surface for the interest of other immune cells, such as T and B lymphocytes. In the case of ingesting a microorganism, these small peptide sequences should be recognised as foreign by the T and B cells that have the appropriate T and B cell receptors on their surface and result in an appropriate immune response (i.e., the digested peptides act as 'antigens'). Ingestion of the body's own cells by a macrophage should not result in an inflammatory immunological response, as any T and B cells recognising 'self-antigens' should have been deleted or inactivated by specific immune control systems. Also, when apoptotic cells are engulfed by macrophages, engagement of the phosphatidylserine receptor stimulates the production of an important signalling molecule, transforming growth factor-β_1 (TGF-β_1), by the macrophage. TGF-β_1 actively suppresses the production of proinflammatory molecules (such as tumour necrosis factor-α) by T cells (i.e., apoptotic cells are proactive in suppressing inflammation). This capability has been tested in animal models of inflammation, which have demonstrated that introduction of apoptotic cells, or of lipid spheres (liposomes) rich in phosphatidylserine, into an area of experimentally induced inflammation can significantly reduce the time required for the inflammation to disappear (resolve). This ability of apoptotic cells clearly has the potential to be exploited for therapeutic benefit.

As already mentioned, in addition to there being mechanisms to recognise the dying, living cells take an active part in rejecting the attention of macrophages. On the surface of lymphocytes there is a protein that goes by the name of CD31. CD31 can bind to other molecules of CD31 on the surface of other cells. The interaction of two cells via their CD31 molecules seems to initiate some kind of active repulsion mechanism. In apoptotic cells this mechanism is faulty and they cannot repel the macrophage. This failure appears to be due to the disabling of one of the components of the downstream signalling pathway from CD31 in the apoptotic cell.

Several other cell surface proteins are also implicated in the recognition and engulfment of apoptotic cells (see Further reading). Some may be general mechanisms and others may be cell- or tissue-specific. But what happens if this all goes wrong and apoptotic cells are left to rot or, more simply, fall apart? As mentioned, a failure to clear apoptotic cells may contribute to the pathogenesis of some diseases. Systemic lupus erythematosis (SLE) is the classic example. This disease is characterised by the production of antibodies to 'self'-proteins (autoantibodies and autoantigens respectively). The autoantibodies that occur in SLE are primarily against components of the cell's nucleus, the nucleosomal proteins, structural proteins around which the cell's DNA is wrapped, and the double-stranded DNA itself. The autoantibody/nucleosome complexes that are formed are frequently deposited within the glomerulus of the kidney and give rise to glomerular nephritis, often associated with SLE. Why are these autoantibodies produced? It seems that it is important for apoptotic cells to be 'eaten whole' prior to falling apart (secondary necrosis). SLE is characterised by the poor clearance of apoptotic cells by macrophages and also decreased immunosuppressive signals from the macrophage following ingestion of apoptotic cells. There is also increased macrophage cell death. Failure to clear apoptotic cells presumably results in increased secondary necrosis and the liberation of cell contents. Why are some of these components not recognised as 'self' and, subsequently, promote an immune response in autoimmune disease? This may have something to do with many proteins being partially cleaved during the process of apoptosis to disable their function. This may generate protein fragments that are not recognised as 'self' and these then perhaps act to directly stimulate the T and B cells. If the cell had been eaten by a macrophage, these protein fragments would have been digested further and consequently would not have appeared as 'nonself'.

Another way that the clearance system for apoptotic cells may be subverted is by microbial pathogens such as *Leishmania amazonensis*. This organism employs a very clever trick of molecular mimicry. By 'looking' like an apoptotic cell, through the expression of phosphatidylserine on its surface, the organism is able to be ingested by a macrophage, where it resides quite happily, resistant to the digestive capabilities of the cell: a biological example of the Trojan horse. The macrophage even helps it further by producing TGF-β_1, thus suppressing any proinflammatory immune response.

In some mammalian tissues, such as the breast epithelium, macrophages do most of the work of removing any apoptotic cells, whereas in contrast, in other tissues such as the gastrointestinal epithelium, it is other neighbouring epithelial cells that rapidly detect and engulf their dying companions or their apoptotic fragments. As a consequence, any of the cell types seen in the intestinal crypts may have phagosomes (cytoplasmic vesicles that result from the ingestion – phagocytosis – of extracellular material) containing the apoptotic remains of a dying neighbour. Even cells in mitosis can be seen with an apoptotic fragments within their cytoplasm. If cell death is extensive, the surviving neighbours may contain several apoptotic bodies (see figure 3). Sometimes these may represent several cell death events spread over a period of time, as recognised by the different degrees of degradation of the apoptotic bodies.

It is clearly important for cells to identify themselves as dying, and it seems that they even go as far as arranging their own funeral. The systems regulating the disposal of dead cells are clearly complex and must operate optimally to avoid serious consequences for the organism.

Further reading

Brown, S., Henisch, I., Ross, E., Shaw, K., Buckley, C. D., and Savill, J. Apoptosis disables CD31-mediated cell detachment from phagocytes promoting binding and engulfmant. *Nature* **418**: 200–203, 2002.

Cohen, P. L., Caricchio, R., Abraham, V., Camenisch, T. D., Jennette, J. C., Roubey, R. A., Earp, H. S., Matsushima, G., and Reap, E. A. Delayed apoptotic cell clearance and lupus-like autoimmunity in mice lacking the c-mer membrane tyrosine kinase. *J. Exp. Med.* **196**: 135–140, 2002.

De Freitas Balanco, J. M., Costa Moreira, M. E., Bonomo, A., Bozza, P. T., Amarante-Mendes, G., Pirmez, C., and Barcinski, M. A. Apoptotic mimicry by an obligate intracellular parasite downregulates macrophage microbicidal activity. *Curr. Biol.* **11**: 1870–1873, 2001.

Dieker, J. W., van der Vlag, J., and Berden, J. H. Triggers for anti-chromatin autoantibody production in SLE. *Lupus* **11**: 856–864, 2002.

Fadok, V. A., Bratton, D. L., Rose, D. M., Pearson, A., Ezekewitz, R. A. B., and Henson, P. M. A receptor for phosphatidylserine-specific clearance of apoptotic cells. *Nature* **405**: 85–90, 2000.

Gregory, C. D. CD14-dependent clearance of apoptotic cells: relevance to the immune system. *Curr. Opin. Immunol.* **12**: 27–34, 2000.

Herrmann, M., Voll, R. E., Zoller, O. M., Hagenhofer, M., Ponner, B. B., and Kalden, J.R. Impaired phagocytosis of apoptotic cell material by monocyte-derived macrophages from patients with systemic lupus erythematosus. *Arthr. Rheumat.* **41**: 1241–1250, 1998.

Huynh, M-L. N., Fadok, V. A., and Henson, P. M. Phosphatidylserine-dependent ingestion of apoptotic cells promotes TGF-β1 secretion and the resolution of inflammation. *J. Clin. Invest.* **109**: 41–50, 2002.

Krieser, R. J., and White, K. Engulfment mechanism of apoptotic cells. *Curr. Opin. Cell Biol.* **14**: 734–738, 2002.

Scott, R. S., McMahon, E. J., Pop, S. M., Reap, E. A. Caricchio, R., Cohen, P. L., Earp, H.S., and Matsushima, G. K. Phagocytosis and clearance of apoptotic cells is mediated by MER. *Nature* **411**: 207–211, 2002.

4

To reproduce or die?

We have already seen in Chapter 2 that one of the most important roles of apoptosis is to act as a counterbalance to cell reproduction. In order to reproduce, a cell must receive not only the right signals to survive (i.e., to ensure that it does not undergo apoptosis) but also the right set of signals and instructions to initiate the process of cell division. Cell division occurs at high levels in many tissues of the body. It is found in cells whose job it is to produce daughters that can act ultimately as the functional cells for the tissue. If apoptosis is to act as a counterbalance in regulating the production of cells in tissues, it is a process that might be expected to be associated at least in part with these reproductive cells.

Defining our terms

In order to fully appreciate the interrelationship between apoptosis and cell division (proliferation) we need to understand the steps and regulations of the cell-reproductive process. This is more complicated in tissues than one might at first suspect and more complicated than is observed in most systems when grown in cell culture media. The first step here is to define a few of the terms. The first of these is *differentiation*. This is the process of cellular specialisation; in our case, changing a proliferative cell into one that is going to perform a specific function associated with that tissue. It is a qualitative term and a relative one. A cell is said to be differentiated relative to other cells in the tissue. All adult cells are differentiated relative to cells in the embryo and many embryonic cells are differentiated relative to the fertilised zygote. Thus you can see that differentiation is associated with a progressive loss of the potential for the cell. A zygote can, and does, make every conceivable cell type that the adult needs and in fact if the two cells resulting from its first division are separated, they will each make two complete embryos (identical twins). In tissues in an adult the dividing cells have a much more restricted potential and are therefore differentiated relative to embryonic cells. However, in several tissues in the adult body a range of differentiation options still exists. Cells in the bone marrow can differentiate into a variety of functional cells (e.g., red blood cells, lymphocytes, granulocytes, and megakaryocytes); however, they are restricted in that they cannot apparently make skin or intestinal cells under normal circumstances, as far as we know at present. Recently though, researchers have demonstrated the ability of bone marrow-derived cells to repopulate and regenerate the liver; maybe the range of options is even greater. Cells in the proliferative compartment of the intestine are capable of making cells whose functional role might be to secrete mucus (the goblet cells), act as absorptive cells (columnar epithelial cells), secrete special hormones (enteroendocrine cells), or play an antibacterial role in terms of their secretion products (Paneth cells). Once the commitment to differentiate into one of these cell types occurs it appears to be an irreversible step. So we can define differentiation as a qualitative change in a characteristic of the cell that is the consequence of the synthesis of one or more novel products, that is, changes in gene expression which lead to the production of new proteins and functional competence of the cell.

Our ability to say whether a cell has undergone differentiation depends on the sensitivity of the techniques that one uses to detect the processes in question. Clearly, a cell that contains fully functional haemoglobin, if it is a red blood cell, or is actively secreting mucus, if it is a goblet cell, can be easily recognised and can be said to be a differentiated functional cell. However, there is a point when the cell is producing its first molecules of haemoglobin, or the first molecules of mucopolysaccharide (mucus). One may also be able to detect these very early initial biochemical products, but even before these are synthesised, messenger RNA (mRNA) has to be made for the cell to construct these molecules and these new mRNAs may be detectable before the protein product is produced. However, even before this, the relevant genes on the DNA have to be read (transcribed) to make the RNA. Now we are close to the limits of our technical ability to detect the differentiation process but there might be yet earlier events related to the derepression or unmasking of the relevant genes.

It is common for some workers in the field to use the expression *terminal differentiation*. It is not always clear what is meant by this term, and it is not always defined by those who use it. Within the context of what was just discussed, terminal differentiation would represent the last step in the sequence of progressive restrictions in the cell's options [i.e., the conversion into a fully functional red blood cell (erythrocyte) or a goblet cell or a surface keratin-containing cell (keratinocyte) in the epidermis of the skin]. Another term that is commonly encountered is *maturation*. Again, this is often used interchangeably with the term *differentiation*, resulting in some confusion and lack of clarity. We suggest that maturation should be thought of as a quantitative change in the cellular constituents associated with the differentiation steps that were just discussed. So a differentiated cell matures as it produces more molecules of mucopolysaccharide or more molecules of keratin.

DNA replication

Proliferation is the term used to describe the cell-reproduction process. In order for a cell to divide there are two essential steps that it has to undergo. Firstly, it has to replicate all its chromosomes, in order that it may give both daughters an equal and complete set. This process is known as DNA replication or DNA synthesis. As mentioned previously, the DNA double helix, consisting of the four phosphorylated DNA bases paired in a precise reciprocating fashion, is coiled and then supercoiled. Unravelling the strands of DNA to enable replication is, therefore, a complex process. Also, any mistakes in the copying would result in a mutation, which may have serious consequences for the cell and the organism as a whole. Careful proofreading is required.

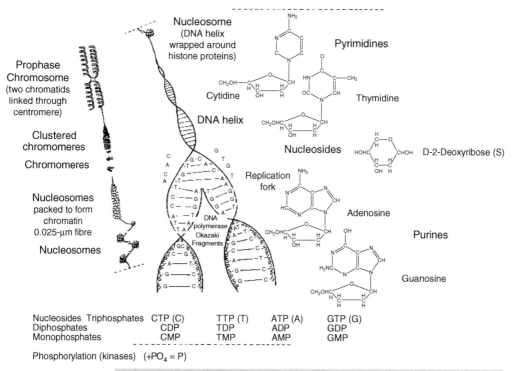

14. Diagram showing chromatin structure and DNA bases (modified from Potten, C. S. Cells. In: Radiation and Skin. Taylor & Francis, London, 1985).

The four DNA bases that form the basis of the genetic code are divided into two classes, *purines*, which are double ringed molecules, and *pyrimidines*, which have a single ring. The purines are found in both RNA (ribonucleic acid) and DNA (deoxyribonucleic acid). They are adenine and guanine (A and G). The pyrimidines found in RNA are cytosine and uracil (C and U). Uracil is similar in structure to thymine (T) the pyrimidine present in DNA. These bases are bound to pentose sugars (D-ribose for RNA) and D-2-deoxyribose for DNA. Once bound to the sugars the molecules are known as *nucleosides* (adenosine, guanosine, cytidine, and uridine or thymidine). Phosphate groups are then added progressively to generate mono-, di-, and then triphosphate molecules known as *nucleotides*. After this phosphorylation the nucleotides can be appropriately paired to make short strands of DNA (known as Okazaki fragments) that can be "stitched" into the DNA helical molecule by an enzyme called ligase. The molecules pair A–T and C–G. This is summarised in figure 14.

So one can see that in order to replicate the DNA, a whole range of complex enzymes and molecules are required. Firstly, when DNA uncoils, it opens up, so that new bases that have been synthesised as a consequence of other enzymes can be incorporated into the replication fork, additional enzymes then join the ends of the short Okazaki fragments and recoil the DNA molecule. Enzymes are required to phosphorylate the *nucleosides* (kinases) and yet other enzymes are required to synthesise the nucleosides. This process has to be performed every time a cell is going to divide and it has to be entirely and successfully completed without errors before the cell itself can divide into two new cells.

Cell division

The second process that is absolutely crucial is the process of cell division or *mitosis*. This again involves a complex series of events to ensure that each daughter cell receives a full complement of chromosomes: forty-six (twenty-three pairs) in humans and forty-eight in mice. Prior to the induction of mitosis, the nucleus consists of a complex ball of DNA material rather like a ball of wool. The individual chromosomes in the ball cannot be recognised. The first stage in mitosis is to condense and contract the DNA material into the discrete chromosomes. There then has to be mechanisms to separate the replicated sets of chromosomes and ensure that one complete set goes into each new daughter cell. Following this, the cytoplasm of the cell has to be divided and new membranes have to be synthesised around the nuclei. Again, this involves a complex programme of events, reliant on the synthesis/degradation and activation/deactivation of specific proteins at precise times.

A specialised form of cell division occurs during the production of the reproductive cells or gametes or germ cells, the egg and the sperm. Here, because these two germ cells are going to fuse at fertilisation, each of these germ cells must contain only half of the DNA material (one maternal and one paternal) so that following fertilisation the cell will again have a full double DNA compliment. This process is known as meiosis and is achieved without the first step of DNA duplication. Meiosis is considered no further; instead, the mitotic cell divisions in adult tissues and those during embryonic development are presented.

Mitosis has been somewhat arbitrarily classified into four distinct phases, based on the appearance of the cell and its chromosomes. The phases are not of equal length and the likelihood of seeing cells at these phases depends on their relative durations. The first phase has been termed *prophase*. It is associated with the condensation from thin strands of DNA into chromosomes. While this is going on, the membrane that surrounds the nucleus is broken down so that the chromosomes will be free to move to their respective daughter nuclei. During prophase, as the chromosomes are condensing, dots and strands of chromosomal material (chromatin) become recognisable and the nuclear membrane gradually disappears. The second phase is known as *metaphase*; this phase is characterised by continued condensation of the chromosomes, which can be identified individually by their size (length) and shape. During metaphase, the chromosomes become aligned in a plate at the centre of what was the nucleus. The chromosomes become attached to contractile molecules of tubulin to form a structure called the mitotic spindle or metaphase spindle. Each tubulin molecule is attached to a particular point on each chromosome known as the *centromere* and when the tubulin contracts, the chromosomes are pulled to the two poles of the spindle. At each pole of the spindle is a special organelle called the *centriole*. This exists as a single structure in proliferating cells that has to be replicated prior to mitosis, with the two centrioles moving to opposite sides of the nucleus and the linking tubulin molecules completing the spindle. When a metaphase cell is viewed face-on, a circular ring of chromosomes may be seen, but viewed from the side a thin line of chromosomes is visible. Of course, every conceivable angle in between these two extremes can be seen and so the appearance of cells at metaphase can differ quite considerably, depending on the angle from which one is viewing the metaphase plate.

The next phase of mitosis is known as *anaphase*. During this phase, the tubulin contracts and draws the two sets of chromosomes to the respective poles. As they are pulled away from the metaphase plate by tubulin the chromosome arms trail and each chromosome tends to adopt a V shape, with the base of the V being the point of attachment (the *centromere*) to the contracting spindle and the two trailing arms of the chromosome making the V. Thus cells undergoing anaphase tend to have a very characteristic appearance, which again depends on the angle from which one is looking at the process. The final phase is known as *telophase*, during which the cytoplasm is split (cytokinesis) and a new nuclear membrane is formed around the two poles at each end of the tubulin spindle. The process is schematically summarised in figure 15.

Proliferation can be regarded as the sequential pattern of changes that occur in gene expression ultimately leading to cell division. It can be simplified to the detection of any of the gene products associated with DNA replication or cell division (mitosis). Because cells tend to go through repetitive rounds of proliferation, the expression of these specialised genes tends to be cyclic.

A particular set of proliferation-associated proteins are known as *cyclins*. The expression of these becomes elevated sequentially as the cell goes through the various processes just described. The pattern of synthesis and expression of cyclins is characteristic of where the cell is in its preparative activity for cell division (see later in this chapter).

The events just described that represent the preparatory sequential processes for cell division can be displayed on a time map that starts at the end of one division and finishes at the end of the next division (mitosis or M). This time map can be expressed as a linear sequence of events or more commonly as a cyclic sequence of events and therefore is referred to as a circular or cyclic map or as the *cell cycle*. This has four major phases or identifiable parts. The first of these, referred to as gap 1 (G1), is the interval between the end of mitosis and the onset of DNA replication. The second phase is the phase of DNA replication known as DNA synthesis phase or S phase. The next phase is the gap between the end of DNA replication and the beginning of the mitotic phase; this is known as gap 2 (G2).

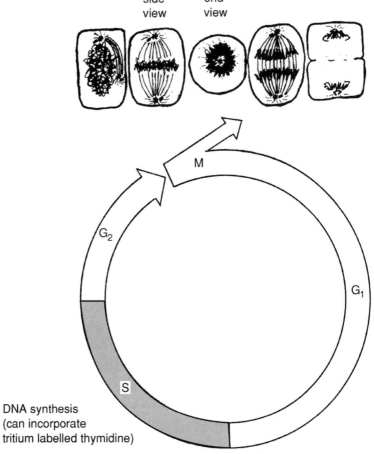

15. Diagram showing the cell cycle and the distinctive appearance of the cell nucleus at the different stages of mitosis (M phase).

The final phase is the process of mitosis itself known as the mitotic phase or M phase. This is shown in figure 15 as the cell cycle. On this time map, a large number of precisely timed, sequential events occur, for example, in association with DNA replication, the timing of the production (synthesis) of each individual sequential enzyme required for DNA synthesis can be recorded, including its point of DNA transcription, its point of messenger RNA production, and its

TO REPRODUCE OR DIE?

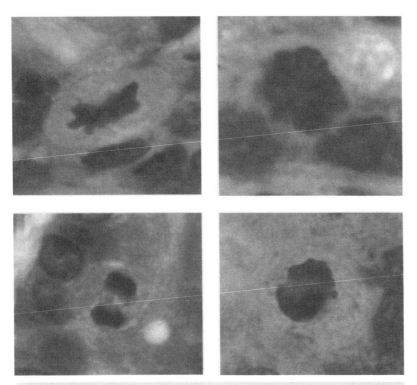

16. Mitotic figures in the small intestinal crypt (haematoxylin and eosin). Top two figures show metaphase plates; bottom two figures show anaphases.

point of synthesis. Because this can in principle be done for each and every enzyme required both in the process of DNA replication and mitosis the map is potentially very complex.

How do we recognise a proliferating cell?

A variety of approaches can be used to determine whether individual cells or populations of cells are proliferating, one of the most obvious is to look and see whether cells within a population or tissue are going through the process of mitosis (figure 16).

Mitosis is associated with condensation of chromatin material into recognisable, individual chromosomes. However, condensation of chromatin is also a feature characteristic of apoptosis and in some circumstances, looking down the microscope, it is difficult to distinguish between cells with condensed chromatin that are at a particular stage in mitosis and cells with condensed chromatin characteristic of apoptosis. In an ideal world this should be simple but in reality there are individual cells in which this distinction is difficult. The distinguishing features that ultimately have to be used are whether the condensed chromatin has sharp boundaries, as seen in apoptosis, and whether the condensed chromatin has any suggestion of chromosome structure (fingerlike, chromosomal protrusions at the edges) that are characteristic of mitosis. One of the difficulties associated with recognising mitotic cells in sections cut through tissues is that the section has a definite thickness and it may incorporate only a small segment of the mitosis. The second problem is, as mention previously, that mitotic cells in metaphase and anaphase have very different appearances depending on the angle at which they are being viewed. As a consequence, a wide range of appearances is associated with mitotic cells in tissue sections. Certainly, if a cell is seen to be in mitosis one can be fairly confident that it is a proliferating cell. In many cases, counting the proportion of cells in a tissue or a cell population that can be seen to be dividing (i.e., in mitosis) gives a measure of the rate of proliferation of that tissue, known as the *mitotoic index*.

Mitosis is a very short phase of the cell cycle and the fraction of cells seen in each phase of the cell cycle is directly proportional to the duration of that phase relative to the whole cell cycle. If the process of mitosis was slowed down, for example, such that it took longer for a cell to complete mitosis, the mitotic index would increase. Alternatively, the mitotic index may increase because the number of proliferating cells in the population being studied increases.

A useful modification of the mitotic index approach is to use drugs that prevent the completion of mitosis. These are agents that tend to stop further progression through mitosis by blocking tubulin synthesis or affecting the mitotic spindle. These drugs are commonly known as spindle poisons. They have been used to kill tumour cells as part of a combination of chemotherapeutic drugs used for treatment of cancer. Examples of these agents are colchicine and colcemid, both of which are derived from a *Crocus*, and vinblastine and vincristine, both of which are derived from the Periwinkle (*Vinca*) plant. A more recent drug is locodazole, which tends to be slightly less toxic and arrests cells for a longer period of time. The approach of blocking cells in mitosis has been termed a *stathmokinetic process*. If cells cannot complete the mitotic phase because the spindle is interfered with, the chromosomes cannot separate at anaphase and cells will stop in metaphase in the presence of these agents. The observation that some cells might be seen in anaphase in such studies is an indication that the action of the drug has either worn off or an insufficient dose was used. The common procedure is to expose cells to the strathmokinetic agents, take samples over a period of time, commonly up to three to four hours and measure the mitotic index. As cells accumulate at metaphase the mitotic index increases with the passage of time. The slope of the line plotted through the accumulating mitotic index gives a measure of the rate of movement of cells into mitosis. Assuming that cells move around the cell cycle at a constant rate, by equating the entry of a cell into mitosis with the rate of exit from mitosis the birth rate of new cells can be determined.

Recognition of cells replicating their DNA

Generally, in adult, mammalian tissues the number of cells replicating their DNA in a proliferating population is approximately seven to fourteen times higher than the number of cells undergoing mitosis because the S phase of the cell cycle is seven to fourteen times longer than the M phase, which in mammalian cells commonly takes 30 minutes to one hour to complete. Thus, one tends to be looking for a much higher proportion of S-phase cells and as a consequence the statistical validity of the data and the ease of obtaining statistically significant numbers is greater. But how do we recognise cells replicating their DNA?

The commonest approach for identifying S-phase cells is to persuade the cells to incorporate DNA bases that are marked or labelled in some way (see figure 14), by incubating the cells in a medium containing 'labelled' thymidine, or by injecting 'labelled' thymidine into an animal. In each case only the cells that are in the S phase of the cell cycle will incorporate the labelled thymidine into their DNA, where it will be permanently located. If the cells are incubated in a

APOPTOSIS: THE LIFE AND DEATH OF CELLS

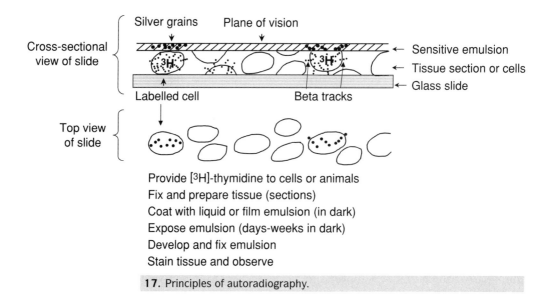

17. Principles of autoradiography.

thymidine-containing medium, they would have to be washed in new nonlabelled thymydine medium to stop the process of incorporation. In the animal, the circulating thymidine is either incorporated into DNA or degraded by passage through the liver within about twenty to thirty minutes, which constitutes a short or pulsed exposure to the labelled material. Longer exposures will result in the labelling of cells as they enter S phase and, therefore, gives no measure of the fraction of cells in the S phase. In principle, any of the four DNA bases could be used, but in practice, mainly because of the specific metabolic pathways involved, thymidine has been the base of choice. Thymidine can be labelled and detected in one of two ways. The first method involves incorporating a radioactive isotope into the molecule and detecting the presence of radioactive decays by coating tissue sections with photographic emulsion, a process called *whole-tissue autoradiography* (or radio autography: see figure 17). The other approach is to incorporate into the thymidine molecule atoms of bromine or iodine, which can be recognised once incorporated into the DNA by antibodies that detect specifically DNA containing bromine- or iodine-labelled molecules using the immunohistochemical techniques described in Chapter 3.

The most common radioactive marker to use is the radioactive isotope of hydrogen called tritium. The thymidine molecule has a methyl group protruding from the side of one of its ring molecules that contains three hydrogen atoms (see figure 14). Tritium is chemically substituted for one of the hydrogens in the methyl group. Other positions in the ring molecule can also be labelled. The molecule could, alternatively, have one of the carbon atoms substituted with the radioactive isotope carbon-14. The advantage of tritium over carbon-14 is that the β-particles emitted by the isotope as it decays are very weak and can penetrate only about 1 micrometre (μm) in tissue, whereas the β-particles from carbon-14 are of higher energy and travel up to ten times further. The photographic emulsion is commonly applied to cells or sections of a tissue on a microscope slide by dipping the slide in a liquid photographic-type of emulsion (earlier studies used thin strips of emulsion). This has to be done in the dark! After dipping, the slides are sealed in a light-tight box and kept (usually in a fridge) for days, or weeks, to allow time for the β-particles released by radioactive decay to accumulate in the emulsion as latent black silver grains. The latent grains can be visualised by developing the photographic film. The silver grains in the developed autoradiograph will be localised over the structures that incorporated the tritiated thymidine (i.e., the replicating chromosomes or the S phase nucleus). Carbon-14 silver grains can spread over more than one nucleus and thus lack resolution. The bromine or iodine atoms have molecular dimensions somewhat similar to those of the methyl group and can be substituted at this point in the molecule. These molecules are known as *bromodeoxyuridine* or *iododeoxyuridene* respectively. The principles of autoradiography are illustrated in figure 17, and an example of a tritiated thymidine autoradiograph and a bromodeoxyuride monoclonal antibody stained section is shown in figure 18.

A variety of sophistications on this technique have evolved with the passage of time that enable other parameters in addition to the fraction of cells in S to be determined. If the cells that have been exposed to tritiated thymidine are allowed to continue to move through the cell cycle (i.e., they are not fixed), then they will enter the mitotic phase and divide into two new cells. The radioactive DNA will

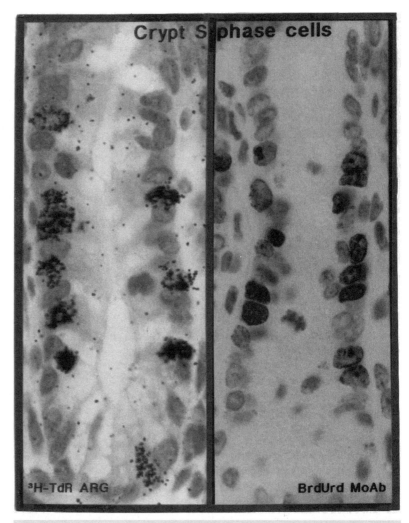

18. Sections through small intestinal crypts showing cells replicating their DNA, either by tritiated thymidine labelling and autoradiography or by bromodeoxyuridine labelling and immunohistochemistry (mitotic cells indicated by arrows).

be segregated in a random fashion between the two new daughter cells, both of which will be labelled. If further time is allowed to pass, the cells will move through G1 into S and then G2 and M (i.e., they will have passed through another cell cycle). With the passage of further time the radioactive label will be diluted further. Such experiments can be used to determine (1) how the number of labelled cells increases with time, which is related to the duration of the cell cycle, (2) how the silver grains (the radioactivity) are diluted with the passage of time, which is also related to the cell cycle, and (3) the speed and timing of the movement of the labelled cells as they pass through mitosis, referred to as the fraction or *percentage of labelled mitosis* (FLM or PLM). This latter technique is very powerful and allows the phases of the cell cycle and the total cycle length to be measured. Other variations on the labelling changes with time are also possible (for example, to look at the formation of pairs and clusters of labelled cells).

In the PLM technique the proportion of radioactively labelled mitotic cells is recorded as time passes from the point of labelling. The PLM initially starts at zero because all the labelled cells will be in the S phase and none in mitosis, but the PLM then rises to 100% as the labelled cells enter the 'mitotic window'. The data generated by this technique tend to be in the form of waves of radioactively labelled mitoses (see figure 19) and the results can be interpreted to provide estimates of the total cell cycle and the duration of the four major phases. Figure 19 shows the theoretical shape of the waves of labelled mitotic cells. In reality the curves tend to be greatly dampened and spread out, which requires computer programmes to fit curves and extract parameters.

Various computer programmes have been devised for analysing such data and providing statistical confidence limits on the parameters measured. The techniques can provide estimates for the total cell-cycle time (T_c), the duration of the S phase (T_s), the duration of $G_2 + M$ (T_{G2+M}), and the duration of G1 (T_{G1}) by subtraction. Bromodeoxyuridine has not been used in grain dilution studies or to any great extent for percentage of labelled mitosis studies. It is more difficult to quantitate differing levels of immunostaining (i.e., brownness; if peroxidase/DAB detection has been used – see Chapter 3) than it is to quantitate a change in the number of black silver grains. Thresholds for identifying a cell as positive are easy to set with autoradiography, but much more difficult for immunohistochemistry.

There are two other antibody approaches that are worth mentioning. The first of these involves an antibody known as *Ki67*. This

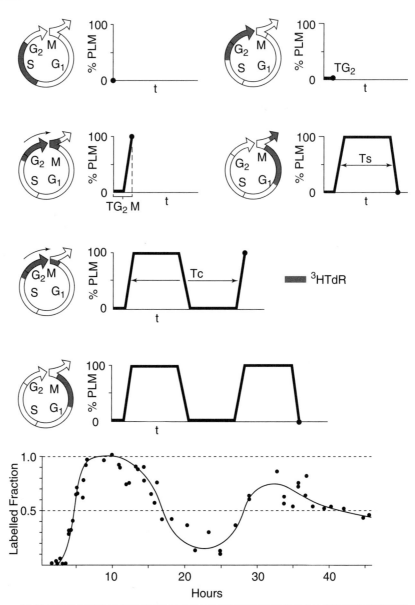

19. The theoretical background to the 'percentage of labelled mitosis' technique for the determination of cell cycle phase durations. The cohort of cells in the S phase are labelled with tritiated thymidine and with the passage of time these cells move through the cell cycle. As they pass through mitosis, the percentage of radioactive mitotic figures rises from 0 to 100 percent. An actual set of data for cells in the small intestinal crypt are shown in the bottom panel.

antibody recognises an antigen expressed in cells that are passing through the cell cycle, the exact nature of which remains somewhat obscure, but the antigen is expressed in cells in most, but not all, phases of the cell cycle. Specifically cells in S, parts of G1, and parts of G2 express this antigen in many cell systems. Recently a monoclonal antibody called *MIB-1* has been developed that recognises the same antigen and this can be used to identify proliferating cells in routinely fixed paraffin sections, unlike *Ki67*, which worked best on frozen sections. MIB-1 is a very versatile and useful probe for proliferating cells. The difficulty one encounters is in interpreting the data that are obtained when one counts the fraction of positive cells because the precise point in the cell cycle where the antigen is first expressed and the point where it ceases to be expressed is not entirely clear and quite possibly varies from cell type to cell type.

A second antibody approach that is also widely used is one that recognises a cyclin-related protein expressed in all proliferating cells. The antigen has been called *proliferating cell nuclear antigen* (*PCNA*) and it is believed to be expressed at all stages of the cell cycle. Staining of proliferating tissue with the available antibodies for PCNA results in essentially all the proliferating cells being positive. The difficulty here is that PCNA seems to be expressed at varying levels in almost all cells (proliferating and differentiated). PCNA is associated with both repair synthesis and regular DNA synthesis and most cells can undergo repair synthesis. The level of PCNA staining observed is very much dependent on the stringency of the immunostaining protocol used; however, if properly standardised this technique can be useful.

Cells can, under appropriate conditions, repair segments of the DNA that have been damaged. This is done by essentially cutting out or excising the segment of DNA that contains the errors. These may be incorrect bases, molecules that are not biochemically correct (i.e., with the wrong atoms bound in the wrong place), or with incorrect types of bonding between atoms. The error-containing fragment is then remade using the error-free segment of complimentary DNA and then stitched back into the DNA molecule by appropriate ligase enzymes. This process of DNA repair synthesis, or, as it is sometimes called, *unscheduled DNA* synthesis, can occur during the G1 or G2 phases of the cell cycle and in differentiated functional cells. The DNA synthesis that is responsible for duplicating the entire DNA complement during S phase is referred to as *scheduled DNA synthesis*. It would be advisable that errors are corrected before scheduled DNA synthesis, or before the segregation of the DNA chromosomes to new daughter cells at mitosis. In addition to the incredible complexity involved in uncoiling the relevant bit of the DNA, cutting out the damaged segment, copying the damaged segment correctly, and reinserting it, the presence of a repair synthetic process indicates that there is also an efficient but complicated system for screening the DNA and detecting DNA damage. The incredibly long molecules of DNA (estimated to be about 3 m in a human cell nucleus) in the chromosomes are constantly being checked in ways that remain unclear.

There is one further complication relating to the cell cycle. For most cells it is assumed that they progress steadily through the sequential biochemical events that represent the cell cycle. Certain conditions may speed the process up or slow it down. The speeding up and slowing down is achieved predominantly by altering the rate of progression through the two gaps in the cell cycle, G1 and G2, particularly G1. For most mammalian cells the length of time that it takes to replicate the DNA and the time taken for mitosis is fairly constant. Typically the S phase duration is about 7 hours (5 to 14 in the extreme), whereas for mitosis it is thirty to sixty minutes. In a mouse, cells in some tissues may pass through the entire cell cycle in 12 hours, whereas others may take 120 hours, ten times longer. Those that take longer to pass through the cell cycle tend to spend long periods of time in G1 or, in some cases, G2. In embryos the cell cycles can be very short.

Cell cycle quiescence

A final consideration that has resulted in considerable debate and argument in the literature over the years is whether cells can stop progression through the cell cycle completely and enter a period of dormancy or quiescence or a phase, to use the cell cycle terminology, of G0 (see figure 20). Some argue that those cells inferred to be in G0 are in fact simply progressing through the G1 phase at an extremely slow rate. This is difficult to prove or disprove but recent work gives clear indications that cells can enter a physiological state of quiescence that is distinguishable from G1 by the absence of the expression of certain proliferation-associated genes, such as *c-myc* and *c-fos*. Thus, cells can enter a state of nonprogression through the cell cycle and it is perhaps during such phases that DNA screening and repair of damage is achieved. These states may exist specifically for such genetic housekeeping. There are indications now that when cells are deliberately damaged, damage-response proteins such as p53 are needed and are involved in the damage recognition and the decision-making process that goes on as to whether a cell is going to repair the damage or commit suicide and die, via apoptosis, by regulating the expression of appropriate sets of genes. If the cell is to repair damage, p53-regulated genes like $p21^{WAF-1/cip-1}$ need to be expressed. This gene codes for a protein that interacts with and inhibits the cyclins that regulate progression through G1 and G2, thus in effect stopping cell cycle progression to allow time for repair.

A distinction should probably be made among the manipulation of cells, for example, in cell culture conditions by altering the temperature, reducing the levels of growth factors, or reducing the quantity of nutrients that can slow down and stop cell cycle progression and the processes outlined above. After such treatment cells are effectively starved and may not be in a physiological G0 state. Similar situations may occur in tumours that may outgrow their blood supply and so some of the tumour cells will find themselves too far from blood capillaries to receive sufficient oxygen and nutrients and as a consequence they will stop cell cycle progression. In culture, cells can divide until the culture dish is crowded and the bottom is covered with a confluent layer of cells. Such confluent cultures tend to stop cell cycle progression and cell division. Normal cells do not pile up and grow one on top of another, which tends to be one of the characteristics of cancer cells in culture. Initially, arrested normal cells still express some proliferation-related gene expression. However, given sufficient time, these genes switch off and they enter a true state of G0. At any time in this arrested state they can be relatively easily stimulated to reenter the cycle and progress further through to S, G_2, and M. In culture this can be achieved by reducing the cellular density, changing the medium, adding growth factors, and so on. In tumours reentry into the cycle can be achieved by increasing the oxygen tension in the tumour or altering the cellular density by, for example, treatment with drugs or radiation to kill off some of the cells near capillaries. Unfortunately the noncycling cells in tumours often tend to be oxygen-deficient and as a consequence are much more resistant to radiotherapy treatment. The cells can become reoxygenated when other cells between them and the blood capillaries are killed and removed. Insufficient studies have been performed to determine whether or not all G0 cells are expressing or not expressing the various proliferation-related genes.

Cyclins

Cyclins are molecules associated with the regulation of the progression of cells through the cell cycle. They form complexes with *cyclin-dependent kinases* (CDKs) or the proteins produced by cell division cycle genes. These complexes are involved with (1) the phosphorylation of the molecules and (2) the activation of the essential substrates required for DNA synthesis and mitosis. The cyclin D family (there are three cyclin Ds) are associated with the movement of cells out of G0 into G1 and with the point of commitment in G1 for undertaking DNA synthesis. This point of commitment, known as a transition or restriction point, has been called 'start'. The complex formed between cyclin-dependent kinase 4 and cyclin D (cdk4-D) lasts for only a short time (i.e., has a short half-life). It forms complexes with molecules like PCNA, protein products of genes like $p21^{WAF-1/cip-1}$, and a tumour suppressor gene known as retinoblastoma (*Rb*). As the cell approaches the onset of the S phase, cyclin E appears and this forms a complex with cyclin-dependent kinase 2 (cdk2-E). This cyclin is degraded during the course of the S phase. Towards the end of the S phase a third form of cyclin appears, cyclin A, which again forms a complex with cyclin-dependent kinase 2 (cdk2-A). This cyclin is expressed at the end of the S phase and in G2. It has also been linked

TO REPRODUCE OR DIE?

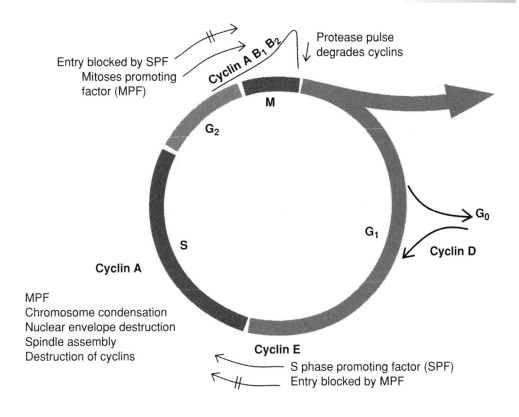

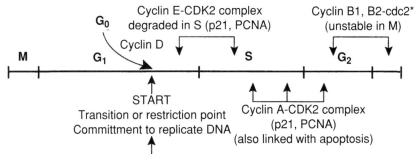

20. Two representations of the cell cycle (a cycle and linear time map) showing the site of action of the cyclins and some of the interacting kinases.

with apoptosis. It may form a complex with cdk2 when the cdk2 is released from cyclin E. Finally, during G2 the B cyclins (there are two of them) are formed. These make a complex with the product of a gene called *cell division cycle* gene 2 (*cdc2*) (which has also been called *mitosis-promoting factor*). This is clearly important for the onset of mitosis. This cyclin is unstable during mitosis and is rapidly broken down. The relationships between the cyclins and the cell cycle are shown in figure 20. The cyclins are clearly important regulators of cell cycle progression. However, subtle differences may occur in the cyclin patterns for different cell systems. The information shown in figure 20 is derived from data obtained from widely diverse systems ranging from yeast to a variety of rather specialised mammalian cell systems.

Flow cytometry techniques

Finally there are flow cytometry-based techniques (see Chapter 2) that enable the fraction of cells undergoing DNA synthesis to be calculated from measurements of the concentration of DNA in cells. The technique is the same as that described for determining apoptosis on the basis of fractional DNA content. Basically, cells that are in the G1 phase will have a single full compliment of DNA, whereas cells that are in the G2 phase will have precisely double this DNA concentration. Cells having a DNA concentration between the G1 and G2 values can be assumed to be replicating their DNA (i.e., be in the S phase). The size and shape of the peaks in DNA concentration for G1 and G2 cells is determined by the relative proportion of cells in these two phases of the cell cycle. Sophisticated computer programmes are available for analysing such data and to extract cell cycle information. Sophisticated approaches have been developed using multiple stains, lasers of different wavelengths, and complex analytical programmes to further enhance these powerful approaches (see figure 10). The techniques can be used in conjunction with antibody labelling, for example, with bromodeoxyuridine antibodies, which can provide further confirmation of the S-phase fraction. As discussed earlier the technique is powerful in that large numbers of cells can be analysed in very short amounts of time, but it is dependent on having suspensions of individual cells. It cannot provide information on spatial interrelationships between cells and between the cells and the tissue. It is also often difficult to discriminate between different cell types, for example, between mature differentiated cells and proliferating cells, both of which would contribute to the G1 peak.

Further reading

Aherne, W. A., Camplejohn, R. S., and Wright, N. A. An Introduction to Cell Population Kinetics. Edward Arnold, London, p. 88, 1977.

Bandura, J. L., and Calvi, B. R. Duplication of the genome in normal and cancer cell cycles. *Cancer Biol. Ther*. **1**: 8–13, 2002.

Baserga, R. Measuring parameters of growth. In: Cell Growth and Apoptosis, G. Studzinski (ed.). IRL Press, Oxford, 1–19, 1995.

Baserga, R. The Biology of Cell Reproduction. Harvard University Press, Cambridge, MA, p. 256, 1985.

Baserga, R. Multiplication and Division in Mammalian Cells. Marcel Dekker, New York, p. 239, 1976.

Blagosklonny, M. V., and Pardee, A. B. The restriction point of the cell cycle. *Cell Cycle* **1**: 103–110, 2002.

Burdon, T., Smith, A., and Savatier, P. Title signalling, cell cycle and pluripotency in embryonic stem cells. *Trends Cell Biol*. **12**: 432–438, 2002.

Cleaver, J. E. Thymidine metabolism and cell kinetics. North-Holland, Amsterdam, p. 259, 1967.

Coqueret, O. Linking cyclins to transcriptional control. *Gene* **299**: 35–55, 2002.

Gardner, R. D., and Burke, D. J. The spindle checkpoint: two transitions, two pathways. *Trends Cell Biol*. **10**: 154–158, 2000.

Hall, P. A., Levison, D. A., and Wright, N. A. Assessment of cell proliferation in clinical, practice. Springer-Verlag, London, p. 210, 1992.

Hunter, T., and Pines, J. Cyclins and cancer II: cyclin D and CDK inhibitors come of age. *Cell* **79**: 573–582, 1994.

Irniger, S. Cyclin destruction in mitosis: a crucial task of Cdc20. *FEBS Lett.* **532**: 7–11, 2002.

Qin, J., and Li, L. Molecular anatomy of the DNA damage and replication checkpoints. *Radiat. Res.* **159**: 139–48, 2003.

Samuel, T., Weber, H. O., and Funk, J. O. Linking DNA damage to cell cycle checkpoints. *Cell Cycle* **1**: 162–168, 2002.

Sherr, C. J. GI phase progression: cycling on cue. *Cell* **79**: 551–555, 1994.

Steel, G. G. Growth kinetics of tumours. Clarendon Press, Oxford, p. 351, 1977.

Tyson, J. J., Csikasz-Nagy, A., and Novak, B. The dynamics of cell cycle regulation. *Bioessays* **24**: 1095–1109, 2002.

Wright, N. A., and Alison, M. The Biology of Epithelial Cell Populations, Vols. 1 and 2. Clarendon, Oxford, p. 1246, 1984.

5

The judge, the jury, and the executioner – the genes that control cell death

p53 – The guardian of the genome in embryos and adults

Here some of the genes that regulate apoptosis are discussed. The gene *p53* codes for a protein that plays an important role in monitoring the genetic fidelity of cells. *p53* is described as a 'tumour suppressor' gene, which means that the p53 protein has a function in preventing the development of cancers. If the gene is mutated, such that an incorrect version of the protein is produced, cancer development may be facilitated. Of the many cancers that have been analysed (more than 2,500 different types of tumour and tumour cell lines), about half of the cancers contain mutations in the *p53* gene, resulting in the absence or dysfunction of the p53 protein. If the p53 gene is transgenically deleted (knocked out) in animals, the animals (termed p53 knockouts or p53 nulls) develop relatively normally; however, tumours (particularly lymphomas) appear in early adult life. Mutations can occur in humans in the germ cell line, which results in a condition known as Li-Fraumeni syndrome; these individuals are born with one normal and one mutant p53 gene and are very prone to cancer. These observations all support the notion that p53 is somehow involved in suppressing the development of cancers.

p53 is an effector protein also involved in the cellular response to DNA damage, transforming signals from proteins that detect radiation or chemical damage to DNA or mistakes in the DNA into changes in gene expression that will allow the cell to deal with the damage. Damage may be dealt with either via repair or via induction

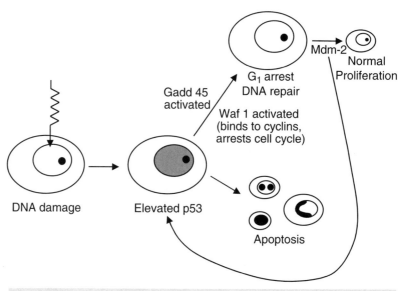

21. *p53*. A simple schematic of the role of p53 protein.

of apoptosis, thus deleting the cell and removing the potential DNA damage, which if left unchecked and unrepaired will ultimately yield a malignantly transformed cell. This hypothesis gave rise to the term *guardian of the genome* as the functional role for p53 (figure 21). p53 has also been shown to be very important in the detection of genetic defects in the early preimplantation and postimplantation mammalian embryo. Under normal circumstances, it is estimated that 50 percent of human conceptions (fertilisations) do not result in implantation in the uterus and are spontaneously aborted. It is believed that p53 induces cells with teratogenic damage (which produces birth defects) to die via apoptosis and the defective embryos to abort. These observations have been strongly supported by experiments where mouse embryos were given a small dose of radiation (2 Gy) on either day $3\frac{1}{2}$ after fertilisation or day $9\frac{1}{2}$. The frequency of birth defects and abortions was recorded together with the levels of apoptosis in embryos that had a normal (wild-type) *p53* gene and animals in which the *p53* gene had been deleted. In the irradiated wild-type animals there was a high level of apoptosis, a high level of abortion, and a relatively low level of genetic abnormalities. In contrast, the levels of apoptosis in the *p53*-null animals were low and

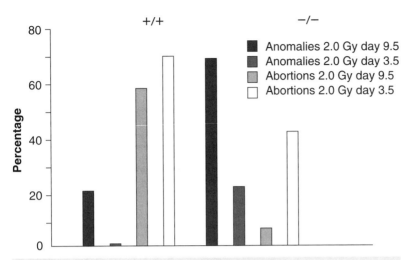

22. Bar chart showing the foetal anomalies and abortions in mice irradiated with 2 Gy on days 3.5 and 9 days of gestation in normal (wild-type +/+) and p53-deficient (p53 knockout, −/−) mice (see also table 4).

unchanged from the unirradiated situation. For animals irradiated at $9\frac{1}{2}$ days, the abortions were low and the abnormalities were very high. The results are summarised in figure 22 and table 4. Therefore, in terms of the life of a mammal, *p53* is crucial in ensuring development of normal embryos and it is possible that the *p53* gene may have developed during evolution to act as a guardian of the genome of the embryo. Recent work suggests that *p53* may be important in other cellular processes, including some related to asymmetric cell division in stem cells.

Therefore, p53 is a checkpoint factor, acting like an emergency brake on the cell cycle so that either repair or apoptosis can be initiated. How p53 is involved in making this decision remains uncertain, but this clearly has profoundly different implications for the fate of the cell depending on which choice is made. To some extent, it is clear that this depends on both cell type and tissue. For instance, proliferative tissues such as the spleen and the intestine demonstrate p53-induced apoptosis in response to ionising radiation-induced DNA damage. In contrast, bone and heart cells do not. Within the intestinal epithelium itself a dichotomy of response exists, with the rapidly proliferating cells of the crypt undergoing apoptosis and the mature,

TABLE 4 Role of p53 in preventing teratogenic effects: study on mice irradicated on day 9.5 with 2 Gy

	p53 ('guardian of the embryo')	
	+/+	−/−
Anomalies (%)	20 (0*)	70 (22*)
Abortions (%)	60 (73*)	7 (44*)
Apoptotic cells (%) 2 Gy on day 9.5 (or 3.5*)	Increased ∼14X	Low & unchanged

P53 induces cells with teratogenic damage to die and defective embryos to abort. Approximately 5 percent of human conceptions not implanted (unrecognisable abortions). Approximately 5 percent of human births have some defects. Norimura et al., *Nat. Med.* **2**: 577, 1996.

nondividing differentiated cells initiating repair. For an individual cell the choice may be governed by the degree to which p53 levels are raised in response to the DNA damage, with low levels of p53 turning on repair genes and higher levels of p53 being required to turn on apoptosis genes or turn off survival genes.

Normally, wild-type p53 is present at extremely low, often-undetectable, levels in most cells. The p53 molecule has a very short biological half-life, which is measured in minutes. This is due to an efficient process that targets the p53 protein for degradation, a process facilitated by a p53 binding protein called mdm2. The levels of wild-type p53 are raised rapidly (within one or two hours) of the induction of DNA damage. It is believed that extremely small levels of damage can be detected and result in the upregulation of the p53 protein, perhaps even a single double-strand break in the DNA might be enough to induce raised levels of p53. Ionising and nonionising (ultraviolet light) radiation induce damage that increases the levels of p53. The levels are also raised following hypoxia (reduced oxygen tension), heating, and cell starvation. The system is sufficiently sensitive such that certain experimental procedures themselves may result in increased p53. For example, the incorporation of tritiated thymidine to label the S-phase cells may be sufficient to cause some DNA strand breaks and therefore increase the levels of p53. p53 levels are probably raised by a combination of molecular stabilisation (prevention of degradation) and increased translation of mRNA transcripts preexisting in the cell. Cells of some tissues such as the brain and skeletal muscle show no p53 response to ionising radiation-induced DNA damage.

p53 positively regulates the expression of a large number of genes, including those coding for p21$^{WAF-1/cip-1}$, mdm2, gadd45, and Bax. p53-induced genes are called PIGS. p53 also represses the transcription of various genes such as those coding for c-fos, c-jun, IL-6, RB (the product of the rentinoblastoma susceptibility gene), and Bcl-2. As mentioned, mdm-2 binds to p53 and targets it for proteolytic degradation and its induction acts as a negative feedback mechanism to limit p53 action. Interestingly, mice that lack mdm-2 die during early embryonic development. *p21$^{WAF-1/cip-1}$* is one of the most studied p53-regulated genes. The p21$^{WAF-1/cip-1}$ protein is a cyclin-dependent kinase inhibitor (CDKI) and is encoded by a gene on chromosome 6p. It binds to and inhibits the phosphorylation and consequent activation of cyclin-dependent kinases, thereby halting cycle progression to induce cell-cycle arrest. The cyclin E/cdk2 and cyclin A/cdk2 are the ones believed to be the most relevant here. In this way p53 may induce arrest of the cells in either G1 or G2, depending on the cell type and repair of DNA damage can then be initiated. Fibroblasts appear to operate a p53-p21$^{WAF-1/cip-1}$ checkpoint in G1, whereas epithelial cells may more commonly operate the p53-p21$^{WAF-1/cip-1}$ checkpoint in G2. p21$^{WAF-1/cip-1}$ also interacts with PCNA to inhibit DNA replication, yet at the same time it may facilitate PCNA's role in DNA repair. gadd45 also interacts with PCNA to inhibit DNA replication.

An alternative option for the cells when p53 is upregulated appears to be to commit cell suicide *via* apoptosis. This may be mediated by p53-dependent expression of *bax*. Bax is a proapoptotic Bcl-2-family protein and functions to promote apoptosis by releasing apoptosis-inducing factors from the mitochondria, an action that is countered by the action of antiapoptotic members of the same protein family, such as Bcl-2 itself (see later in this chapter). If Bcl-2 expression is increased this can block the p53-mediated apoptosis.

The main experimental evidence strongly linking the p53 to DNA damage-induced apoptosis is the fact that in *p53* knockout animals,

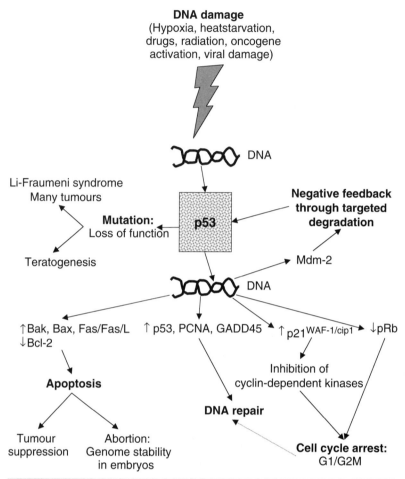

23. A chart showing the central role for p53 in determining the cellular response to DNA damage and some of the genes and molecules that interact with it.

radiation does not induce the apoptosis seen in normal animals. The various interactions and pivotal role of p53 are illustrated in figures 21 and 23. In humans, the *p53* gene is located on chromosome 17p and there are various hot spots in the gene for mutation. Many mutations in the genetic code lead to changes in the amino acid sequence of the protein (missense muations) and as a consequence may affect the function of the protein. For instance, p53 must bind to DNA to regulate gene transcription; therefore, changes in amino acid sequence within the region of the protein responsible for

DNA binding may block its transcriptional effects. Almost 80 percent of all *p53* missense mutations described so far in human tumours are within the region coding for the DNA binding portion of the molecule. Mutated p53 is found in a large number of cancers and in the germ line of Li-Fraumeni patients. However, it should be pointed out that the mutation of the *p53* gene alone is not sufficient to explain the cancer incidence figures and it would appear that several other mutations are also required. In the Li-Fraumeni patients, a *p53* mutation preexists. Even though these patients have one good copy of the p53 gene, their ability to increase p53 levels in response to DNA damage is highly compromised and over the period of the next ten to thirty years these patients accumulate the other important mutations and eventually develop cancers. In the more common cancer situations in non-Li-Fraumeni individuals (i.e., the normal population), the other mutations may accumulate with increasing age, with mutations in the *p53* gene often being one of the later changes to be observed in tumours as they become more malignant.

In skin exposed to ultraviolet (UV) light, a known skin carcinogen, *p53* mutations are found in precancerous lesions and in more than 90% of human squamous cell carcinomas. Following exposure to UV, cells in skin may die and generate cells with a changed appearance known as sunburn cells, which are now thought to represent the epidermal equivalent of apoptosis. When *p53* is deleted, mice exposed to UV have reduced numbers of sunburn cells.

Extensive studies on apoptosis have been conducted on lymphocytes cultured for varying periods of time. Lymphocytes from adult animals are not generally active from a proliferative point of view but can be stimulated into proliferation in a variety of ways. Lymphocytes derived from animals where the *p53* gene was deleted do not respond by apoptotic death as is seen when the *p53* gene is fully functional. Similar findings have also been reported for the epithelial cells in the intestine. However, if the lymphocytes were first stimulated to proliferate then they could die *via* apoptosis even though no *p53* gene was present. This suggests that there are both *p53*-dependent and *p53*-independent pathways for apoptosis.

Genes that determine survival or death – the *bcl-2* family

Whether a cell survives to proliferate or differentiate, or whether it dies *via* apoptosis, is to some extent determined by a whole series of genes producing factors (proteins) that instruct the cells to live or die. The ultimate fate of the cell depends on the balance between these survival and death proteins. During the course of evolution, many of these genes are conserved; that is, there is considerable similarity in the DNA coding and protein composition across an enormously wide range of organisms from nematode worms, flies, amphibians, mammals, and humans. Also, during the course of evolution a wide range of variants have developed in terms of the genes and their proteins so that mammalian cells have not just a single gene determining survival and a single gene determining death, but a whole family of survival and death genes. A given pair of these genes may be the predominant functional members, but the other genes can act as reserves and come into play if the major genes are damaged (mutated) or deleted through the natural occasional loss of a chromosomal segment or by transgenic manipulation in the experimental situation. This redundancy of similar and closely related genes provides a powerful reserve and protection for the cells. p53 acts somewhat like a judge on the system, determining the fate of a cell, whereas all the survival and death genes perhaps are acting like a jury and directing the judge and the executioner.

The B-cell lymphoma 2 gene, or *bcl-2* for short, was the first of these survival death genes to be identified. Patients with non-Hodgkin's disease, a form of B-cell lymphoma, have a particular chromosomal translocation that results in the juxtapositioning of the *bcl-2* gene (from chromosome 18), with the immunoglobulin heavy chain gene locus (on chromosome 14). Immunoglobulin genes are highly expressed in B cells and following the translocation, t(14:18), so is the *bcl-2* gene. The *bcl-2* gene was sequenced and cloned and several research groups showed that by inserting the *bcl-2* gene into cells they could protect them against a range of stimuli that could induce apoptosis.

Bcl-2 protein has a molecular weight of 25–26 kDa. It has a membrane-anchoring domain at its carboxyl terminus and it is found on the cytoplasmic face of intracellular organelles, such as the mitochondria, endoplasmic reticulum, and the nucleus. Its function appears to be to regulate ion flux across the organellar membranes, thereby contributing to the maintenance of electrochemical gradients in the cell.

Bcl-2 expression is seen in a variety of tissues throughout the body. Lymphoid tissues such as the spleen, thymus, tonsils, and Peyer's patches of the small intestine show strong expression. Expression is also seen in the epithelial cells that line the ducts in the breast and in the epithelial cells at the very base of the colonic crypts, the colonic epithelial stem cells. In contrast, Bcl-2 expression is not observed in the epithelial cells of the small intestinal crypts, at least in our studies. The significance of this observation is discussed in Chapter 7. All the tissues just mentioned are characterised by high levels of cell proliferation and apoptosis (i.e., cell turnover).

After the characterisation of Bcl-2, the search was initiated for proteins that could bind to Bcl-2. The first to be discovered was called Bax, Bcl-2-associated X protein, a proapoptotic molecule that antagonised the effects of Bcl-2. Also discovered at this time was another homologue, Bcl-x. The *bcl-x* gene was found to have two mRNA transcripts, a long form coding for an antiapoptotic molecule ($Bcl-x_L$) and a short form coding for a proapoptotic molecule ($Bcl-x_S$). The fact that Bax could bind to either $Bcl-x_L$ or Bcl-2 (heterodimerise) and that all three proteins could bind to themselves (homodimerise) led to the idea that a complex equilibrium between the homo- and heterodimers of these pro- and antiapoptotic molecules determined whether a cell lived or died. Many homologues of Bcl-2, both pro- and antiapoptotic, have now been discovered. They share regions that have very similar amino acid sequences (conserved structural domains), which are simply called Bcl-2 homology domains or BH domains (figure 24). Four of these domains have been identified. The BH1 and BH2 domains are involved in the dimerisation of Bcl-2 family proteins. The BH3 domain is essential for the death-promoting properties of the

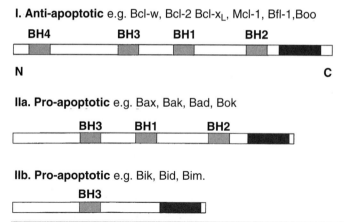

24. Figure showing different Bcl-2 homlogy domains that are possessed by different Bcl-2 family members.

proapoptotic proteins such as Bax, although antiapoptotic Bcl-2 family proteins also have this domain. Some proapoptotic members of the family, such as the protein Bid, possess only the BH3 domain. Finally, the BH4 domain is found only in antiapoptotic Bcl-2 family proteins and is required for their interaction with non-Bcl-2 family proteins, including the protein kinase Raf-1 and the calcium-dependent phosphatase calcineurin.

Therefore, what are the precise mechanisms that are involved in the regulation of apoptosis by Bcl-2 family proteins? Studies with the nematode worm *C. elegans* have greatly facilitated our understanding of Bcl-2 function and apoptosis in general. The importance of the worm in this and other areas of cell biology is clearly demonstrated by the 2002 award of the Nobel prize to researchers who pioneered the use of the worm as an experimental model. The worm possesses an evolutionarily conserved homologue of *bcl-2* called *ced-9*: ced is an acronym of C. elegans death gene. During its lifetime the worm produces 1090 cells, of which 131 are deleted by apoptosis; these are mostly neurons deleted during modelling of the animal's nervous system in embryogenesis, leaving an adult worm with exactly 959 cells. The deletion of these 131 cells is dependent on two other *ced* genes, *ced-3* and *ced-4*: Both *ced-3* and *ced-4* are obligatory for apoptosis in *C. elegans*. Experiments showed that the role of *ced-9* was to counter the effects of these two genes. *ced-3* codes for a protease and *ced-4*

TABLE 5 Mammalian homologues of C. elegans genes

C. elegans	Mammalian
CED-3	Caspase (9)
CED-4	Apaf-1
CED-9	Bcl-2
EGL-1	Bax

for a ced-3 binding protein. The binding of the ced-3 protein to ced-4 facilitates self-activation of the ced-3 protease; this is blocked when ced-9 is also bound to ced-4. Apoptosis in the worm may be initiated by displacement of ced-9 from the complex with ced-3/ced-4 by another protein, egl-1 (egl stands for 'egg laying defective', a phenotype of worms that lack a normal *egl-1* gene). Egl-1 is the *C. elegans* homologue of the proapoptotic protein Bax and interestingly was one *C. elegans* gene that was discovered (some time) after its mammalian homologue was characterised (figure 25 and table 5).

With the discovery that ced-3 was a protease, a search for mammalian homologues was initiated. The first homologue was found to be a known enzyme, interleukin-1β-converting enzyme, or ICE for short. Since then, fourteen different homologues have been discovered. A few years ago they were renamed and given a specific nomenclature. These proteases are termed caspases: cysteine proteases that cleave after aspartic acid residues. ICE is now called caspase-1 and the true mammalian homologue of ced-3 is caspase-9.

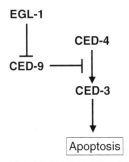

25. Cell death pathway in *C. elegans*.

Caspases are now known to cleave many cellular proteins during apoptosis and are responsible for many of the classical features of an apoptotic cell. Their role in apoptosis is discussed more fully later.

So, in the worm ced-9 blocks caspase activation by binding to the ced-4 protein. Does Bcl-2 have an equivalent function in mammals? It took a while longer to find a mammalian homologue for ced-4 than it did for ced-3, but eventually one was found. It was given the name Apaf-1, apoptotic protease activating factor 1. More recently a ced-4 homologue was found in the fruit fly; it is called DARK. A whole family of related proteins called CARD (caspase recruitment domain) proteins have now been characterised. Experiments have demonstrated that Apaf-1 can bind ced-3 and caspase-9 and that Bcl-x_L can bind to ced-4, but as yet it has not been shown that any mammalian Bcl-2 family member can bind to Apaf-1.

Bcl-2 and Bcl-x_L can block caspase-9 activation, but how? In order for Apaf-1 to facilitate the activation of caspase-9, it needs to dimerise, thus bringing two molecules of caspase-9 into juxtaposition. This process is dependent on ATP and the binding of an accessory factor by Apaf-1. This factor is cytochrome c, a component of the mitochondrial respiratory chain. Apaf-1 binds cytochrome c through specific amino acid motifs at its carboxyl terminus, called WD-40 repeats.

Release of cytochrome c is due to the transitory opening of voltage-sensitive pores in the outer mitochondrial membrane. These pores (voltage-dependent anion channels: VDAC) open in response to a decrease in the electrical potential difference between the cytoplasm and the mitochondrial intermembrane space (depolarisation) induced by insertion of Bax into the outer membrane. Depolarisation is induced as Bax molecules form pores through which positively charged ions may move from the cytoplasm, down their electrochemical gradient, and into the more negatively charged mitochondria. Cytochrome c may exit via the open VDACs and perhaps via the Bax pores. Other proapoptotic molecules, including apoptosis-inducing factor (AIF) and endonuclease G, are also released during mitochondrial membrane depolarisation. Cytochrome c release is blocked by Bcl-2 and Bcl-x_L. The action of Bcl-2 and Bcl-x_L may be dependent on interaction with non Bcl-2 family proteins and be independent of proapoptotic Bcl-2 family members; alternatively, it may depend on heterodimerisation with Bax or its close functional relative, Bak. It is likely that both mechanisms are involved.

Although there are several antiapoptotic Bcl-2 family proteins, they are not functionally equivalent and if one is absent its loss may not be compensated for by the actions of the other proteins. This is highlighted in studies on mice in which different family members have been knocked out. Bcl-2-null mice (i.e., no Bcl-2 expression) are runts, much smaller than their littermates that have either one or two copies of the *bcl-2* gene. These mice also can develop a polycystic kidney disease (a rare, inherited condition in humans). The cysts form due to the inappropriate, large-scale apoptosis of kidney cells. The masses of dead cells calcify, forming cysts, and impair kidney function. In contrast, $Bcl-x_L$-null mice die *in utero*, approximately two thirds of the way through their foetal development. These mice show many defects, especially in the brain. Bcl-2 and Bcl-xL clearly have different roles then when viewed in the context of the whole organism.

So *p53* may be the molecule that judges whether a cell is going to repair damage or die and bcl-2 and its family of genes provide information that helps determine the fate, whereas the protease enzymes act as the executioner of the cell.

Apoptotic proteases

During the course of evolution a number of protease enzymes have evolved that are responsible for breaking down various products in the cell and have also developed a range of other functions. As discussed above, the *C. elegans* death gene, *ced-3*, was identified as coding for a cysteine protease and subsequently the first mammalian homologue of this protease was discovered. This was interleukin-1β (IL-1β) converting enzyme, or ICE for short, which acts to cleave the plasma membrane-bound form of the proinflammatory cytokine, IL-1β, thus releasing it to act as a secretory factor. Fourteen different mammalian enzymes with homology to ICE have now been discovered. These are

TABLE 6

Caspase subfamily	Caspase	Function	Functional domains
IL-1β converting enzyme/caspase-1 related caspases	1 4 5 11 12 13	Inflammation: processing of cytokines Apoptosis (some systems?)	CARD-containing caspases
CED-3-related caspases	2 9	Activated by intrinsic 'death signals'	
	8 10	Activated by extrinsic 'death signals'	DED-containing caspases
	3 6 7	Effector caspases activated by caspases-2,-9,-8 and -10	Short prodomain – no self-association and autocatalytic activation

now termed *caspases*, a term we defined earlier. Caspase enzymes have also been identified in the fly, in amphibians, and in plants. The caspase family is summarised in table 6.

So, what do they do? Well, their obvious function is to cleave cellular proteins. They effectively deconstruct the cell. The wide range of proteins currently recognised as being substrates for caspases is listed in table 7. Many of these proteins are structural and their cleavage allows the cellular condensation. Also disabled are many proteins involved in execution of the cell cycle and effecting DNA repair; in addition, a specific DNase is activated by caspase-mediated cleavage of a DNase-inhibitory protein that is normally bound to the DNase enzyme. This DNase goes by the obvious name of caspase-activated DNase, CAD.

Caenorhabditis elegans has clearly shown us that the ced-3 protease is required for apoptosis; however, studies in mice have shown that its mammalian homologue caspase-9 is not essential for normal foetal development – a process that is dependent on the deletion of unwanted or damaged cells by apoptosis. There is, therefore, some degree of functional redundancy in the caspase family of enzymes. This family of enzymes can be broadly subdivided into two groups

TABLE 7 Examples of Cellular Caspase Substrates

Protein classification	Substrates
Cytoplasmic and nuclear structural proteins	G-actin
	Fodrin
	Gelsolin
	Lamins A and B
Nuclear proteins involved in DNA repair and control of cell cycle	DNA-protein kinase (DNA-PK)
	Ku
	MdM-2
	Retinoblastoma protein (pRB)
	Poly(ADP-ribose) polymerase
	p27^{kip1}
Signal transduction	Focal Adhesion Kinase (FAK)
	Protein Kinase C-γ (PKC-γ)
	Receptor Interacting Protein (RIP)
Endonuclease inhibitors	Inhibitor of caspase-activated DNase (ICAD)

based on their ability to initiate self-activation (autocatalysis) and to activate other caspases. Caspase-9 is an initiator caspase: it is capable of self-activation when complexed with Apaf-1 and cytochrome c – this complex has been dubbed the 'apoptosome'. Other initiator caspases of similar type to caspase-9 include caspases 1, 2, 4, 5, and 11. These caspases are not recruited by the Apaf-1 protein and activated in response to cytochrome c release, but they can bind to other Apaf-1-like or CARD proteins (see earlier) and may be activated by other cellular stress signals, for example, caspase 1 binds to the protein CARD 12. There is a second set of initiator caspases, which are characterised by caspases 8 and 10. These caspases are not activated in response to internal signals, like cytochrome c release but are activated by extracellular signals. These signals are either membrane-bound or secretory polypeptides that engage cell surface receptors to initiate a sequence of events that result in caspase activation. The receptors are members of the tumour necrosis factor receptor (TNF-R) superfamily. The members of this family that can initiate apoptosis are called death receptors (DRs).

The first 'death receptor' to be characterised was Fas (aka Apo-1/CD 95); its ligand is called FasL. FasL is expressed strongly on T lymphocytes and natural killer cells that have been activated in response to an appropriate immune stimulus. Fas receptor expression may be increased in cells in response to viral infection. Cytotoxic agents may also upregulate Fas expression in a *p53*-dependent manner. When a cell expressing Fas on its surface comes into contact with the activated T lymphocytes or natural killer cells it may be targeted for deletion by apoptosis. So, the Fas/FasL system is primarily associated with immunological surveillance and destruction of cells that are hazardous to the host. It is also involved in limiting the number of proinflammatory cells during an immune response.

The other most-studied members of the family are the receptor–ligand pairs TNF-R1 and TNF-α and DR5 and TNF-R-apoptosis-inducing ligand (TRAIL). When the ligand binds to its DR, it induces the trimerisation of the receptor (binding of three molecules together). The receptors have a conserved cytoplasmic domain called the death domain (DD). Via this domain they can bind other DD proteins that act as adapter proteins for binding caspases-8 and/or -10. The receptor Fas requires one adapter, called Fas-associated death domain protein (FADD), in order to recruit caspases, whereas TNF-R1 also needs to bind TNF-R-associated death domain protein (TRADD) before it can bind FADD. FADD binds caspases by another conserved domain, the death effecter domain (DED). It is probable that the receptor, adapter(s), and caspases all exist in a complex together, prior to ligand binding. The ligand merely induces trimerisation, thus bringing caspase molecules into close apposition with each other and facilitating autocatalysis. The entire complex of proteins is called the death-inducing signalling complex, or DISC.

There are some serine proteases, like Granzyme B, that have a similarity to caspases. Granzyme B is an enzyme that is produced by cytotoxic T-cell lymphocytes; these are a subset of T lymphocytes that can target and kill cells that express the antigen they specifically recognise (an antigen may be defined as any molecule, or part of a molecule, to which the body can mount an immune response: The specificity of the cytotoxic T cell will be defined by the specific and

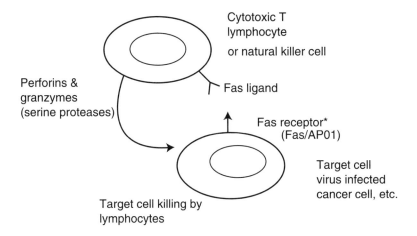

* Member of Tumour Necrosis Factor (TNF) receptor family.

26. Cytotoxic T-cell-induced killing may be mediated by both the Fas/FasL pathway and perforin/granzyme B to activate caspase-dependent apoptosis in target cells.

unique T-cell receptor it expresses on its surface). Cytotoxic T cells inject the Granzyme B into the target cell, through a porelike protein (perforin) that they insert into the target cell plasma membrane. Granzyme B is similar to caspase-8 in its preferred sequence of amino acids in that it binds to and cleaves, and it activates the same downstream signalling events as caspase 8 to induce apoptosis of the target (see figure 26).

The presence of many caspases with overlapping expression patterns suggests that functional redundancy between the various members exists, unlike in the Bcl-2 family. Mice that lack individual caspase enzymes develop normally and have few deficiencies. Mice deficient in caspase-1, for example, are quite normal in appearance and behaviour; however, they are resistant to the toxic effects of bacterial endotoxins. Mice deficient in caspase-9 are also normal.

The big picture

It is clear that both healthy cells and cells that have been exposed to a damaging agent have genetic and molecular programmes to receive instructions (signals) and to respond to these signals. This response

APOPTOSIS: THE LIFE AND DEATH OF CELLS

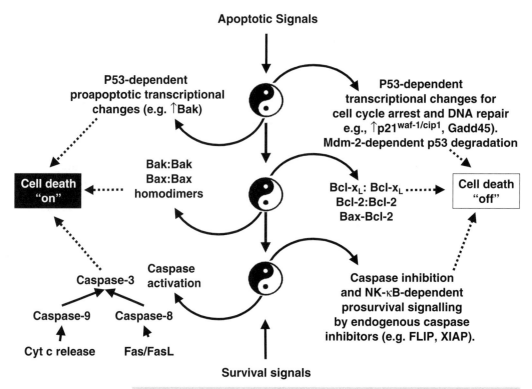

27. Interaction of p53, Bcl-2 family proteins, and caspases in determining cell life or death.

may be the sequence of events that lead to apoptosis or alternatively cell cycle arrest and repair. If apoptosis is initiated, further regulatory processes act at the level of the tissue to ensure that the damaged cells are removed and that their loss is compensated for. This is clearly quite a complex cell and tissue programming mechanism (see figure 6).

This chapter should have allowed appreciation of the fact that apoptosis is governed by multimember protein families, such as the Bcl-2 family, the death receptors (TNF-R superfamily), CARD proteins, and the caspase enzymes. The role of *p53* in genome surveillance and how a cell may determine its own fate following genotoxic insult (figure 27) were also discussed. Also described was how both internal and external stimuli may initiate cell death, initially through independent pathways, but ultimately through common effectors, such as caspase-3 and endonucleases (figure 28).

THE JUDGE, THE JURY, AND THE EXECUTIONER

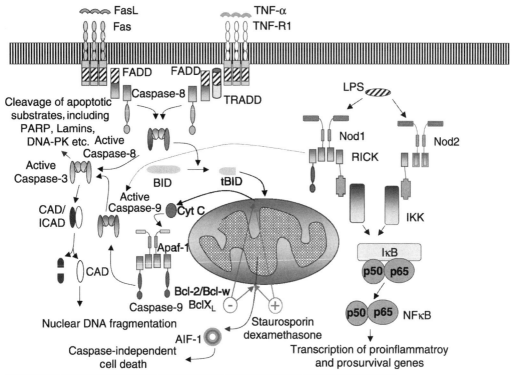

28. Diagram showing both signalling pathways for induction of cell death via death receptors and via stimuli that result in mitochondrial membrane permeability transition. The interaction of the two pathways should be noted.

The following chapters discuss some of these topics with specific relation to the intestine, so that they can be appreciated more fully within organ systems of whole animals.

Further reading

Bakkenist, C. J., and Kastan, M. B. DNA damage activates ATM through intermolecular autophosphorylation and dimer dissociation. *Nature* **421**: 499–506, 2003.

Bartek, J., and Lukas, J. Damage alert. *Nature* **421**: 486–488, 2003.

Boise, L. H., Gonzalez-Garcia, M., Postema, C. E., Ding, L., Lindsten, T., Turka, L. A., Mao, X., Nunez, G., and Thompson, C. B. *bcl-x*, a *bcl-2*-related gene that functions as a dominant regulator of apoptotic cell death. Cell **74**: 1–20, 1993.

Chen, M. W. Initiator caspases in apoptosis signaling pathways. *Apoptosis* **7**: 313–319, 2002.

Chen, X., Ko, L. J., Jayaraman, L., and Prives, C. p53 levels, functional domains, and DNA damage determine the extent of the apoptotic response of tumor cells. *Genes Dev.* **10**: 2438–2451, 1996.

Ferri, K. F., and Kroemer, G. Organelle-specific initiation of cell death pathways. *Nature Cell Biol.* **3**: E255–E263, 2001.

Gagliardini, F., Fernandez, P. A., Lee, R. K., Drexler, H. C., Rotello, R. J., Fishman, M. C., and Yuan, J. Prevention of vertebrate neuronal death by the crmA gene. *Science* **263**: 826–828, 1994.

Gregory, C. D. Apoptosis in the immune system. In: Apoptosis in Normal Development and Cancer, M. Sluyser (ed.). Taylor & Francis, London, pp. 223–352, 1996.

Grossmann, J., Artinger, M., Grasso, A. W., Kung, H.-J., Schölmerich, J., Fiocchi, C., and Levine, A. D. Hierarchical cleavage of focal adhesion kinase by caspases alters signal transduction during apoptosis of intestinal epithelial cells. *Gastroenterology* **120**: 79–88, 2001.

Hendry, J. H., Broadbent, D. A., Roberts, S. A., and Potten, C. S. Effects of deficiency in p53 or bcl-2 on the sensitivity of clonogenic cells in the small intestine to low dose-rate irradiation. *Int. J. Rad. Biol.* **76**: 559–565, 2000.

Hockenberry, D., Nunez, G., Milliman, C., Schreiber, R. D., and Korsmeyer, S. J. Bcl-2 is an inner mitochondrial membrane protein that blocks programmed cell death. *Nature* **348**: 334–336, 1990.

Hockenbery, D. M., Zutter, M., Hicker, W., Nahm, M., and Korsmeyer, S. Bcl-2 protein is topographically restricted in tissues characterised by apoptosic cell death. *Proc. Nat. Acad. Sci. USA* **88**: 6961–6965, 1991.

Hoyes, K. P., Cai, W. B., Potten, C. S., and Hendry, J. H. Effect of bcl-2 deficiency on the radiation response of clonogenic cells in small and large intestine, bone marrow and testis. *Int. J. Rad. Biol.* **76**: 1435–1442, 2000.

Inohara, N., Koseki, T., del Pseo, L., Hu, Y., Yee, C., Chen, S., Carrio, R., Merino, J., Liu, D., Ni, J., and Nunez, G. Nod-1, an Apaf-1-like activator of Caspase-9 and Nuclear Factor-κB. *J. Biol. Chem.* **274**: 14560–14567, 1999.

Joza, N., Susin, S. A., Daugas, E., Stanford, W. L., Cho, S. K., Li, C. Y. J., Sasaki, T., Elia, A. J., Cheng, H.-Y. M., Ravagnan, L., Ferri, K. F., Zamzami, N., Wakeham, A., Hakem, R., Yoshida, H., Kong, Y.-Y., Mak, T. W., Zuniga-Pflucker, J. C., Kroemer, G., and Penninger, J. M. Essential role of the mitochondrial apoptosis-inducing factor in programmed cell death. Nature **410**: 549–554, 2001.

Kastan, M. B. Onyinye, O. Sidransky, D., Vogelstein, B., and Craig, R. W. participation of p53 protein in the cellular response to DNA damage. Cancer Res. **51**: 6304–6311, 1991.

Kastan, M. B., Zhan, Q., El-Deiry, W. S., Carrier, F., Jacks, T., Walsh, W. V., Plunkett, B. S., Vogelstein, B., and Fornace, Jr., A. J. A mammalian cell cycle checkpoint pathway utilising p53 and *GADD45* is defective in Ataxia-Telangeictasia. Cell **71**: 587–597, 1992.

Kaufmann, S. H., Desnoyers, S., Ottaviano, Y., Davidson, N. E., and Poirier, G. G. Specific proteolytic cleavage of Poly(ADP-ribose) polymerase: an early market of chemotherapy-induced apoptosis. Cancer Res. **53**: 3976–3985, 1993.

Kuida, K., Haydar, T. F., Kuan, C.-Y., Gu, Y., Taya, C., Karasuyama, H., Su, M. S.-S., Rakic, P., and Flavell, R. A. Reduced apoptosis and cytochrome c-mediated caspase activation in mice lacking caspase 9. Cell **94**: 325–337, 1998.

Lane, D. P., p53, guardian of the genome. Nature **358**: 15–16, 1992.

Lane, D. P., and Lain, S. Therapeutic exploitation of the **p53** pathway. Trends Mol. Med. **8**: S38–42, 2002

Lemasters, J. J., Nieminen, A-L., Qian, T., Trost, L. C., Elmore, S. P., Nishimura, Y., Crowe, R. A., Cascio, W. E., Bradham, C. A., Brenner, D. A., and Herman, B. The mitochondrial permeability transition in cell death: a common mechanism in necrosis, apoptosis and autophagy. Biochim. Biophys. Acta **1366**: 177–196, 1998.

Li, K., Li, Y., Shelton, J. M., Richardson, J. A., Spencer, E., Chen, Z. J., Wang, X., and Williams, R. S. Cytochrome *c* deficiency causes embryonic lethality and attenuates stress-induced apoptosis. Cell **101**: 389–399, 2000.

Luo, X., Budihardjo, I., Zou, H., Slaughter, C., and Wang, X. Bid, a Bcl-2 interacting protein, mediates cytochrome c release from mitochondria in response to activation of cell surface death receptors. Cell. **94**: 481–490, 1998.

Lu, Q.-L., Poulson, R., Wong, L., and Horby, A. M. Bcl-2 expression in adult and embryonic non-haematopoietic tissues. *J. Pathol.* **169**: 431–437, 1993.

Marshman, E., Ottewell, P. D., Potten, C. S., and Watson, A. J. Caspase activation during spontaneous and radiation-induced apoptosis in the murine intestine. *J. Pathol.* **195**: 285–292, 2001.

Martin, S. J. Dealing the CARDs between life and death. Trends Cell Biol. **11**: 188–189, 2001.

Mayo, L. D., and Donner, D. B. The PTEN, Mdm2, p53 tumor suppressor-oncoprotein network. Trends Biochem. Sci. **27**: 462–467, 2002.

Merritt, A. J., Potten, C. S., Watson, A. J. M., Loh, D. Y., and Hickman, J. A. Differential expression of bcl-2 in intestinal epithelia: correlation with attenuation of apoptosis in colonic crypts and the incidence of colonic neoplasia. J. Cell Sci. **108**: 226–271, 1995.

Merritt, A. J., Allen, T. D., Potten, C. S., and Hickman, J. A. Apoptosis in small intestinal epithelial from p53-null mice: evidence for a delayed, p53-independent G2/M-associated cell death after gamma-irradiation. Oncogene, **14**: 2759–2766, 1997.

Merritt, A. J., Potten, C. S., Kemp, C. J., Hickman, J. A., Balmain, A., Lane, D. P., and Hall, P. A. The role of p53 in spontaneous and radiation-induced apoptosis in the gastrointestinal tract of normal and p53-deficient mice. Cancer Res. **54**: 614–617, 1994.

Michael, D., and Oren, M. The p53 and Mdm2 families in cancer. Curr. Opin. Gen. Dev. **12**: 53–9, 2002.

Nagata S. Apoptosis by death factor. Cell **88**: 355–365, 1997.

Nordstrom, W., and Abrams, J. M. Guardian ancestry: fly *p53* and damage-inducible apoptosis. Cell Death Different. **7**: 1035–1038, 2000.

Norimura, T., Nomoto, S., Katsuki, M., Gondo, Y., and Kondo, S. *p53*-Dependent apoptosis suppresses radiation-induced teratogenesis. Nat. Med. **2**: 577–580, 1996.

Ogura, Y., Inohara, N., Benito, A., Chen, F. F., Yamaoka, S., and Nunez, G. Nod-2, a Nod1/Apaf-1 family member that is restricted to monocytes and activates NF-κB. J. Biol. Chem. **276**: 4812–4818, 2001.

Oltavi, Z. N., Milliman, C. L., and Korsmeyer, S. J. Bcl-2 heterodimerises in vivo with a conserved homolog, Bax, that accelerates programmed cell death. Cell **74**: 609–619, 1993.

Oren, M. Damalas, A. Gottlieb, T., Michael, D., Taplick, J., Leal, J. F., Maya, R., Moas, M., Seger, R., Taya, Y., and Ben-Ze'ev, A. Regulation of *p53*: intricate loops and delicate balances. Biochem. Pharmacol. **64**: 865–871, 2002.

Reed, J. C. Bcl-2 and the regulation of programmed cell death in cancer. In: Apoptosis in Normal Development and Cancer, M. Sluyser (ed.). Taylor & Francis, London, pp. 127–169, 1996.

Rodriguez, A., Oliver, H., Zou, H., Chen, P., Wang, X., and Abrams, J. M. Dark is a *Drosophila* homologue of Apaf-1/CED-4 and functions in an evolutionarily conserved death pathway. Nat. Cell Biol. **1**: 272–279, 1999.

Sharpless, N. E., and DePinho, R. A. p53: good cop/bad cop. Cell **110**: 9–12, 2002.

Shi, Y. A structural view of mitochondria-mediated apoptosis. Nat. Struct. Biol. **8**: 394–401, 2001.

Spector, M. S., Desnoyers, S., Hoeppner, D. J., and Hengartner, M. O. Interaction between *C. elegans* dell-death regulators CED-9 and CED-4. Nature **385**: 653–656, 1997.

Susin, S. A., Lorenzo, H. K., Zamzani, N., Marzo, I., Snow, B. E., Brothers, G. M., Mangion, J., Jacotot, E., Constantini, P., Loeffler, M., Larochette, N., Goodlett, D. R., Aebersold, R., Siderovski, D. P., Penninger, J. M., and Kroemer, G. Molecular characterisation of mitochondrial apoptosis-inducing factor. Nature **397**: 441–446, 1999.

Tewari, M., Quan, L. T., O'Rourke, K., Desnoyers, S., Zeng, Z., Beidler, D. R., Poirier, G. G., Salvesen, G. S., and Dixit, V. M. Yama/CPP32β, a mammalian homolog of CED-3, is a CrmA-inhibitable protease that cleaves the death substrate poly(ADP-ribose) polymerase. Cell **81**: 801–809, 1995.

Thornberry, N. A., and Lazebnik, Y. Caspases: enemies within. Science. **281**: 1312–1316, 1998.

Vaux, D. L., Weissman, I. L., and Kim, S. K. Prevention of programmed cell death in *Caenorhabditis elegans* by human bcl-2. Science **258**: 1955–1957, 1992.

Watson, A. J., Merritt, A. J., Jones, L. S., Askew, J. N., Anderson, E., Becciolini, A., Balzi, M., Potten, C. S., and Hickman, J. A. Evidence of reciprocity of bcl-2 and p53 expression in human colorectal adenomas and carcinomas. Br. J. Cancer **73**: 889–895. 1996.

Wilson, J. W., Booth, C., and Potten, C. S. (eds.). Apoptosis Genes. Kluwer, Boston, p. 310, 1998.

Xue, D., and Horvitz, H. R. *Caenorhabditis elegans* CED-9 protein is a bifunctional cell-death inhibitor. Nature **390**: 305–308, 1997.

Yuan, J. Y., and Horvitz, H. R. The *Caenorhabditis elegans* genes *ced-3* and *ced-4* act cell autonomously to cause programmed cell death. Dev. Biol. **138**: 33–41, 1990.

Zou, H., Henzel, W. J., Liu, X., Lutschg, A., and Wang, X. Apaf-1, a human protein homologous to C. elegans CED-4, participates in cytochrome c-dependent activation of caspase-3. Cell **90**: 405–413, 1997.

6

Stem cells

What is a stem cell?

So far the simple situation where tissues are divided into proliferating cells that are passing through the cell cycle, and differentiated functional cells that are no longer capable of cell division has been presented. Differentiated mature cells have finished with cell cycle activity and have synthesised the proteins associated with their functional role. Here, I wish to raise the question of whether all proliferative cells are equal in terms of their proliferative potential; whether they all perform the same reproductive function, and whether it might be possible that some cells early in the differentiation pathway might still be capable of passing through a few cell cycles.

If we consider dividing cells a number of possibilities exist as can be seen in figure 29. This diagram considers individual cells but it is also possible to consider populations of cells in a similar way and in fact, it is probably more correct to do so. In the simplest theoretical situation in 29A, a cell divides producing two daughters, both of which become functional differentiated cells. In this case, the original proliferating cell does of course disappear and if we consider the original cell, its proliferative potential was very restricted (to only one division). If all the proliferative cells in a tissue behaved like this, cell production would quickly cease and the tissue would shrink away, so this is unlikely to occur. Nevertheless, occasional individual proliferative cells within a population of proliferating cells might behave like this. Figure 29B is a simple modification of 29A, where the original cell at the bottom of the diagram divides and produces two daughters.

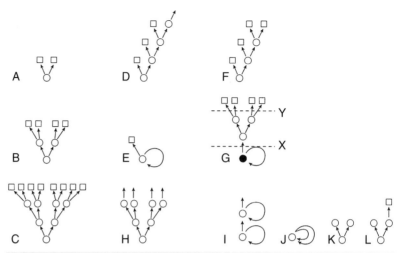

29. A range of possible cell lineage diagrams to explain cell replacement in renewing tissues.

The original cell again disappears from the picture when it divides, but its two daughters continue to progress through another cell cycle and divide again producing four differentiated functional cells from the original cell but again after these two cycles all the proliferative cells will have disappeared. However, the original proliferating cell at the bottom of the diagram can be considered to have a larger division potential than the proliferative cell in 29A. This line of reasoning can be continued for any number of divisions in this amplifying type of scheme (as shown in figure 29C); however, in all these cases the proliferating cells eventually disappear, because none of the daughter cells have the ability to maintain their own numbers. So, these are not schemes that can sustain a tissue over a period of time, as in all of them the proliferating cells are not self-maintaining. Figure 29D is a version of 29A, where the proliferative cell is exhibiting self-maintenance at each division, because it produces another proliferating cell like itself that can go through the cell cycle again and again, each time producing a differentiated cell. This type of division scheme can continue *ad infinitum* and can be redrawn as shown in figure 29E. The major difference between figures 29D or 29E and 29A is that 29A has what can be regarded as a symmetric division process; both daughters are identical. Whereas figures 29D and 29E show asymmetric divisions where the daughters differ, one being proliferative (a circle) and the other differentiated (a square). In figure 29H, 29I, 29J, and 29K, symmetrical divisions are represented where only self-maintaining proliferating cells are produced. Here, the tissue would expand rapidly, in fact exponentially, but no differentiated functional cells would be produced. So this is also an unrealistic scenario for the replacing tissues of the body. Such schemes may exist early in embryological development when there is a rapid expansion in cell number and all the cells are essentially equal in their potential for future divisions and differentiation, and also in some cell cultures. This type of symmetric division producing only identical proliferative daughters may occur amongst occasional proliferative cells in a stable tissue, but if it does the frequency with which it occurs must equal the frequency of the symmetric type of division shown in figure 29A so that, on average, in all the cells a situation is achieved that equates to that in figure 29E. If this balance between differentiated cells and proliferative cells is not equal, as shown in 29E, the tissue will either grow or shrink depending on whether the proliferative or the differentiated cells are dominant. This imbalance, resulting in expansion of the proliferative compartment, clearly occurs during development, in tissue growth after injury (wound healing), and in tissue growth in cancers, and possibly the reverse is seen during the shrinkage of tissues associated with ageing. Thus, in adult tissues under conditions of stable tissue size, a state that we refer to as *steady-state conditions*, the situation depicted in 29E must, on average, exist. The proliferative cells here are self-maintaining (i.e., their numbers remain constant). This self-maintenance process can be thought of in terms of the probability of the proliferative cells producing other proliferative cells, or the *self-maintenance probability*. Clearly under steady-state conditions this must be equal to 0.5. If this probability is raised, more proliferative cells than differentiated cells will be produced, whereas if it reduces below 0.5 for any significant length of time the tissue (number of cells) will shrink – a situation seen in old age in some tissues and referred to as hypoplasia or aplasia.

An asymmetric lineage, such as that depicted in 29D, could in theory terminate at some point in time, as shown in figure 29F, in which case the proliferative cells are lost from the system because all are converted ultimately to differentiated cells. This is really only

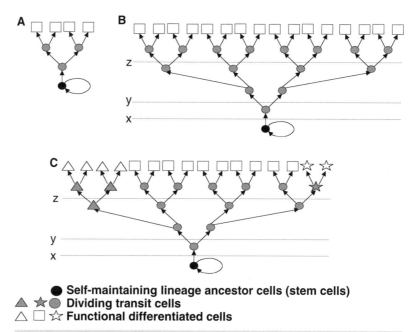

30. Lineage diagrams showing stem cells and dividing transit cells and the different levels of differentiation that separate stem cells from dividing transit cells (see also figure 29G).

a modification of that scheme shown in figure 29A and can occur in tissues providing it is compensated for by the types of divisions depicted in figures 29H, 29I, 30J, and 30K to ensure that on average the situation in figure 29E is achieved.

These diagrams illustrate two other important points; firstly, as already stated it is better to think of these processes in terms of average processes for a large number of cells and not to consider them for individual cells. Secondly, in order to say anything about the division potential of a cell one has to be able to either know what the cell has done in the past or design an experiment so that you can see what it is going to do in the future in terms of cell divisions. For example, looking at figure 29F, over a short time scale one would see asymmetric divisions but looking over a longer time scale, it would be a symmetric process with extinction of the proliferating cells.

So on average, in steady-state conditions, in adult tissues of a mouse or a human, the situation depicted in figure 29E must be achieved. There is one further theoretical consideration that we need to take into account. The situation depicted in 29E, could be achieved if every time a proliferating cell divides it produces one daughter indistinguishable from the mother cell and a second daughter that has undergone a differentiation step. In this situation all the proliferating cells are essentially identical, they all perform the same proliferative role, and they all have the same large proliferative potential. It is unclear, and remains a matter of some debate, whether the divisions here are truly asymmetric (i.e., whether one daughter differs from the other from the moment of birth) or whether the situation in figure 29E is achieved *via* a symmetric division in the first instance with both daughters produced being proliferative (figure 29K) but one is then exposed to differentiation signals and rapidly removed as shown in figure 29L. Although these are somewhat academic considerations they do have implications in terms of the mechanisms operating at the molecular, cellular, and tissue levels. There is a school of thought that argues that all divisions amongst the proliferating cells are of the type illustrated in figures 29K and 29L and that it is the local environment (niche), cell position, or polarity of the divisions that determines whether one of the cells becomes a differentiated cell.

Figure 29E and the comments in the preceding paragraph imply that all proliferating cells are equal but there is an alternative theoretical possibility as illustrated in figure 29G. Here we have a proliferating cell, which is self-maintaining (closed circle) at the origin of an amplifying scheme. The initial proliferating cell produces a daughter that then passes through a division sequence identical to that in figure 29B, eventually producing four differentiated daughters. In doing so the three proliferating cells shown in the middle region of the figure (between X and Y) disappear and become differentiated cells.

As in figure 29B, this part of the diagram would not be self-maintaining if it were not for the ancestral cell at the base of the scheme, which has divided to produce an input for the middle part of the diagram. This type of scheme we describe as an *hierarchical* or *lineage* based scheme. It is characterised by at least two types of proliferating cells: *self-maintaining lineage ancestor cells* or *stem cells*, as shown at the bottom of the diagram, and other proliferating cells that have a limited division potential, which are referred to as *dividing transit cells*. This is not the place, or time, to go into all the arguments but it is now quite widely accepted that most of the tissues of the body that exhibit cell division activity (the so-called replacing tissues) have a proliferative organisation of the sort described in figure 29G.

Those tissues of the body that exhibit constant proliferation are the tissues where functional cells are constantly being worn out and lost and therefore need replacing. These tissues are known as *renewal* or *replacing* tissues of the body and examples are the skin, the lining of the intestine, the blood-forming tissues in the bone, certain glandular tissues, and sperm-producing cells in the testes. In all these cases the evidence now suggests cellular hierarchies or lineages of the sort depicted in figures 29G and 30. What differs amongst these various tissues is the number of divisions that occur in the dividing transit population as shown in figure 30. In schemes such as that shown in figures 29 or 30 the simplest assumption to make is that a differentiation event occurs at the time of division of the lineage ancestor cell (as depicted by the dotted line X in figures 29G and 30B). However, differentiation could in principle be delayed until Y or even Z in figure 30B. If differentiation is delayed, then in effect all the cells below the dotted line depicting the differentiation event are identical to the lineage ancestor cell and the fact that they are drawn as part of a dividing transit compartment is because the tissue architecture or structure determines that usually these cells are moving or in transit through the tissue. This will be explained in greater detail in the next chapter when we look at one renewing tissue, the intestine.

Stem cell definition

Stem cell is the name that has been given to a self-maintaining, lineage ancestor cell. The topic of stem cells is one that raises much discussion in the scientific literature and many definitions exist. Part of the reason for the many definitions is that the concept of a stem cell is somewhat dependent on the scientific context within which stem cells are being considered. Individuals working with invertebrate systems (worms and flies), those working with mammalian embryos or plants, those working with glandular tissue, steady-state renewing tissues like skin, gut, and bone marrow, and those working with regenerating wounded tissues or even tumours all tend to have a slightly different interpretations of what a stem cell is. However, a few generalised conclusions can be drawn. We have already presented one of the simplest definitions, namely that a stem cell is the self-maintaining ancestral cell of a lineage or hierarchy. This implies that it has a large, perhaps continual, division potential. However, there are a few additional considerations that can be added to the definition and it is perhaps slightly more correct when talking about adult renewing tissue stem cells to say that they are relatively undifferentiated cells, capable of extensive proliferation with self-maintenance and that they may produce many different differentiated progeny. If the tissue is injured, the stem cells, because they are the cells with the greatest division potential, are probably the most efficient cells at healing and regenerating the damaged or wounded tissue. They appear to be cells that are capable of a fairly flexible use of all these different options. Because they are, by definition, the only cells in a tissue that persist throughout adult life and because cancer development is known to take many years (up to decades in humans) and requires several genetic changes (mutations) in the cell, cancer is thought by many to be a disease of stem cells. However, this is questioned by some.

There are a few further points to consider. In figure 29E, the production of a differentiated cell effectively removes a cell from the proliferative compartment. Thus when considering the proliferative compartment, differentiation is equivalent to cell loss or death. We have also seen that the probability of a cell differentiating or of dividing further must be equal (at 0.5) in steady state for the stem cells. If the probability of proliferation (self-maintenance) is tipped in favour of proliferation, then the tissue will grow as is seen in a cancer. A cancer can develop if the self-maintenance probability is only slightly raised (e.g., from 0.5 to 0.51). An important point to bear in mind is that the removal of a cell from the proliferative compartment can equally well be achieved by either cell death (apoptosis) or differentiation, so the balance between cell division and whether a tissue grows, shrinks, or remains stable is determined by the sum of the cells that differentiate and those that die via apoptosis relative to the number of new cells produced within the proliferative compartment by cell division (mitosis) (see table 8). One current view is that it may be

TABLE 8 Pertubations of Steady State		
Apoptosis plus differentiation	→	Cell Loss
Proliferation	→	Cell production
Cells lost + cells produced At each division	→	Steady state (stable tissue) 50%^{a}/50%^{b}
If cell loss increases relative to proliferation	→	tissue shrinks (ageing?) 51%^{a}/49%^{b}
If cell loss decreases relative to proliferation	→	tissue expands (cancer?) 49%^{a}/51%^{b}
If proliferation increases relative to cell loss	→	tissue expands (cancer?) 49%^{a}/51%^{b}
If proliferation decreases relative to cell loss	→	tissue shrinks (ageing?) 51%^{a}/49%^{b}

^{a} Percentage of new born cells lost to apoptosis or differentiation.
^{b} Percentage of newborn cells that remain as proliferative cells.

apoptosis that is changed during carcinogenesis with slightly fewer stem cells dying via apoptosis than is normal and as a consequence the proliferative compartment gradually expands.

Most of the renewing or replacing tissues of the body have an hierarchical scheme, or cell lineage, such as that shown in figure 29G or in figure 30, and such schemes are characterised by the majority of the proliferating cells being dividing transit cells. These cells vastly outnumber the self-maintaining stem cells and they typically are transient residents in the tissue. In the bone marrow, the stem cells may constitute only about a fraction of the 1 percent of the proliferating marrow cells, whereas in the epithelial tissues they may constitute between 1 and 10 percent (see figure 31). They are also transitory cells and often move physically as they develop from the proliferative compartment to the mature functional compartment where they become senescent and are shed or removed. In some tissues such as the intestine their life expectancy from the moment of birth from a cell division of a stem cell may only be a few days.

Cancers often take several years in a mouse, and decades in humans, to develop from the time of exposure to a carcinogen or the induction of the initial carcinogenic change in the cell, to the formation of a cancer. Consequently, it is argued, that it is unlikely that cancers arise in the dividing transit cells and it can be concluded that cancer is a problem associated with the regulatory processes determining self-maintenance and cell loss (asymmetric versus symmetric cell

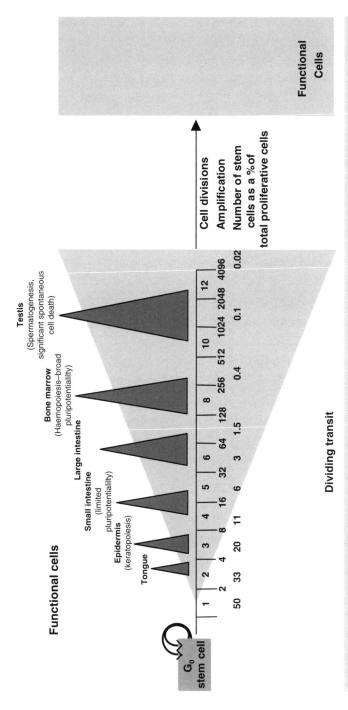

31. Diagram showing the cell lineages in the various replacing tissues of the body with the size (number of cell divisions) of the dividing transit population. Also shown is the consequent amplification in cell numbers generated by the dividing transit cell population and the inverse proportion of stem cells.

division) in the stem cell compartment. It is for this reason that the identification of stem cells and an understanding of their regulation is a topic of considerable interest.

In both the bone marrow and the intestine, there are some indications that the stem cell compartment itself may be structured as a short lineage, or hierarchy as is suggested by the figure 30B if you consider that the point of commitment to differentiation is the dashed line Z rather than X.

One further consideration relates to the differentiation events. Figure 30B shows essentially one type of differentiation event, the production of squares. However, cells may, and often do in tissues, differentiate down different pathways (figure 30C). This can occur at any level in the lineage and can result in the production of several different functional cell types, as is seen in the bone marrow and intestinal cell lineages. If this happens, the ancestral stem cell can be said to be pluripotent, in terms of differentiation options. Embryo stem cells (at least the ealy embryo cells) tend to be totipotent, because they can make all types of cell.

It is currently a matter of considerable debate and clinical interest as to whether adult tissue stem cells retain a greater totipotency than was previously thought or could be experimentally demonstrated. It is possible that given the correct signals, adult tissue stem cells could make all the differentiated cells types of the body. This debate centres around the issue of plasticity (moldability) of the stem cells. The only clear example is that under some circumstances adult bone marrow cells seem to be capable of making liver cells. However, it should be remembered that the foetal liver is an active blood-forming organ like the adult bone marrow. It is also clear that in certain abnormal situations, some of which are thought to be precancerous, patterns of abnormal differentiation (metaplasia) can be seen in tissue.

Stem cells and tissue injury

If a tissue that is hierarchical is exposed to agents that kill cells, including stem cells, the tissue must be regenerated and the stem cell population reestablished, if the tissue itself is not to shrink in size and disappear. In reality, damaged tissues usually shrink initially and are then regenerated and it is believed that this is achieved by regenerative cell proliferation in the surviving stem cell population. If regenerating stem cells are to make new stem cells to replace those killed, they must adopt, at least transiently, some symmetric divisions of the sort illustrated in figures 29H to 29K. In reality, if they adopted a policy of only having symmetric divisions then the stem cell population would expand rapidly in size but no differentiated cells would be produced which would be to the detriment of the function of the tissue. Consequently, what appears to happen in most incidences is that the self-maintenance probability changes from the 0.5 that characterises steady-state stable growth (figure 29E) (i.e., asymmetric divisions) to something greater than 0.5 but less than 1.0 (which would be the value of self-maintenance in purely symmetric expansionary growth). This ensures that some differentiated cells are produced while the stem cell population is being reestablished. Thus the long-term maintenance of the functional integrity of the tissue (e.g., whether the skin or intestinal epithelium develop holes or ulcers) depends on the number of stem cells that survive and the competition between the consequences of stem cell death and regenerative stem cell divisions. The regenerative divisions repopulate the stem cell population and also produce some differentiated cells, which may continue to divide as they pass through the cell lineage. An hierarchical system provides certain advantages and flexibility in relation to this competitive race. Both the number of divisions and the speed with which cells move through the divisions in the dividing transit population can be regulated. The speed with which the stem cells go through the cell cycle and their self-maintenance probability can also be regulated. These determine the number of cells per unit time that enter into the bottom of the hierarchy and so enter the dividing transit population.

Another potential advantage of hierarchies is that the stem cell population need be only very small, so there are few cells in the tissue at risk of suffering genetic damage, developing cancers, and so on. These stem cells also only need to divide relatively infrequently to maintain the structure. Most of the cell division work and cell production is achieved by the dividing transit population. Fewer cell cycles in a lifetime reduces mutation risk (cancer risk) as does long cell cycles that allow plenty of time for repair of genetic damage.

Self-maintenance probability

We have already touched on the topic of self-maintenance: the ability of a population of cells to maintain its own numbers. It is not a concept that is particularly useful when considering individual cells within schemes such as that shown in figure 29. However, the proliferative cell in figure 29A has a self-maintenance probability of zero, whereas the proliferative cell in figure 29J has a self-maintenance probability of 1.0, and under steady-state conditions as exemplified by figure 29E, the self-maintenance probability must be 0.5. Self-maintenance probability should really be thought of in terms of populations of cells, and the probability of self-maintenance is then the probability that cells within a population produce daughters like themselves. It would only require a very small increase in the self-maintenance probability of stem cells to generate over a long period of time a cancerous like growth (see table 8). The self-maintenance probability (sometimes represented as p_{sm}) is influenced by the size of the population of cells that is considered and/or the length of time over which it is considered.

A test of functional competence for stem cells: clonogenic cells

Unfortunately stem cells generally have no distinguishing features, because they are defined in part as unfifferentiated cells; neither are there any particular stains or markers such as monoclonal antibodies that one can use to identify them, with the possible exception of the bone marrow stem cells. They cannot be recognised in either tissues or in suspensions of cells prepared from tissues; however, there are various modern developments that might soon enable stem cells to be marked and studied. These involve various antibodies to cell surface, or intracellular, structures. The lack of markers has made it very difficult to study these all-important cells.

The most widely used approach for studying properties associated with stem cells has been to take advantage of the fact that these are the cells that will be the most efficient regenerators of the tissue following injury. Tests have been devised for many of the renewing tissues that were initially developed to study tissue regeneration in the bone marrow. A landmark technical article was presented in 1961 where the regeneration of the haematopoietic (blood forming) tissues was studied in animals that had been irradiated with doses that killed many or most of the bone marrow stem cells. The few surviving stem cells, or more commonly bone marrow transplanted stem cells, circulated around the body and lodged in various sites, including the spleen and bone marrow, where they began to regenerate both the stem cell population and the extensive haemopoietic transit cell lineages. The technique was more effective when it involved transplantation of appropriately treated bone marrow into the blood of animals that had first been irradiated with a dose of radiation that completely sterilised their own bone marrow stem cells. The transplanted bone marrow stem cells seeded into sites in the body favourable for blood cell regeneration (haemopoiesis) and one of these, the spleen, has been extensively studied. The stem cells that seeded into the spleen divided rapidly and formed lumps of growing bone marrow tissue. In spleens that were appropriately fixed these regenerating colonies could be observed and counted with the naked eye. This technique became known as the spleen colony assay or the *colony forming units in spleen* (CFU-S) assay. The cells that formed these colonies were bone marrow stem cells and each colony could be shown to develop from a single stem cell. It is now realised that CFU-s are not derived from the ultimate stem cells in the bone marrow but are formed of early progenitor cells (i.e., are part of an hierarchical stem cell lineage). An enormous variety of ingenious experiments were designed that enabled the numbers of bone marrow stem cells to be estimated, their cell cycle status determined, their response to cytotoxic insults, and their responsiveness to growth factors studied. Because each colony could be shown with genetic markers to be derived from an individual surviving stem cell, the colonies can be regarded as clones and the cells that form these colonies are called *colony forming cells*, *clone forming cells*, or *clonogenic cells*. Over the next twenty years a considerable insight into the cellular hierarchies of the murine haemopoietic tissue was obtained using this test of functional stem cell potential.

The bone marrow studies have evolved considerably with the realisation that the colony forming cells could be grown in culture, and the culture techniques have thrown further light onto the nature of the stem cell population and the cellular hierarchies, or age structure, that occur therein. Gradually, similar approaches were adopted to study stem cells in the epidermis of skin and in the crypts in the epithelium or mucosa that lines the intestinal tract. The approach has been adapted to study stem cells in the testis and with less efficiency and scientific rigor in various other tissues. To date, extensive studies on clonal regeneration have been undertaken in the intestine, which has led to a considerable insight into the lineage organisation and the lineage ancestral stem cells in this tissue.

There are, however, one or two problems associated with these clonogenic regenerative assays. Firstly, it is an indirect approach; the stem cells are not studied directly. They are looked at from the point of view of what they are capable of doing under rather extreme conditions. So it is a test of their future potential under extreme conditions rather than their actual functional properties in steady state. This has led some to think of these studies as suffering similar limitations in some respects to *Heisenberg's uncertainty principle* in quantum physics. This states, in grossly simplified terms, that to study a particular entity like an electron, the experiment changes the nature of the entity, leading to an uncertainty in the measurements that are obtained. Here, in order to study stem cells we have to expose the cells and tissues to extreme damage and this may alter their characteristics, including those that are being measured.

A wide range of systems has been studied using these approaches, often involving different environmental conditions for the stem cells compared with their natural habitat. They may be performed *in situ* following severe injury, they may involve transplantation or grafting, which results in the stem cells being located in sites where they are normally not found, or they often involve cell culture experiments (an even more unnatural environment). One of the major differences amongst the different techniques that have been used has been the adoption of different criteria for identifying the colonies. The colonies in spleen, and those observed *in situ* in the gut and skin, use the most stringent criteria for identification. Here the colonies are large and may contain several thousands of cells, which requires a very large number of cell divisions from the initial clonogenic cell. However, in other systems the size of the colonies may be very small or the number of divisions that are implied for the survival of a clonogenic cell is very limited. If this is the case it inevitably raises the question as to whether stem cells are really being studied because some cells in the early transit lineage might be capable of producing colonies of a limited and small size (i.e., they may be quite capable of one, two, or a few regenerative type cell divisions).

Are stem cells intrinsically different from transit cells?

One final complication with the concept of stem cells that has not been fully resolved concerns the extent to which stem cells are intrinsically different from the other proliferating cells in the dividing transit population. Are they intrinsically different or are they merely instructed by other cells, or the environment, to behave differently? This clearly relates to the questions concerning symmetrical and asymmetric divisions. If stem cells are not intrinsically different, then any differences seen must be imposed from outside the cell (i.e., from the microenvironment), which means that the stem cell divisions were symmetric. The other possibility is that the stem cells differ intrinsically from transit cells but no one has identified the intrinsic properties. The former possibility, that stem cells are simply proliferative cells that are instructed to function as lineage ancestor cells by the environment in which they exist, suggests that if they were placed somewhere else (in a different environment) they would behave differently. Similarly, if other proliferative cells were placed in the stem cells environment or niche, they would behave like stem cells. This is one of those questions that have not been answered by appropriate studies, but in fact it is extremely difficult to design the appropriate experiments that would convincingly solve the problem. This issue becomes particularly pertinent when considering the clonogenic stem cells and when interpreting the results of clonal regeneration experiments. It appears that cells one or two cell generations (divisions) away from the ultimate stem cell can regenerate the stem cell compartment and the tissue if all the ultimate stem cells are killed or reproductively sterilised. This suggests that they have not, at this stage, undertaken an irreversible commitment to differentiation. It is argued that as a consequence they are indistinguishable from the ultimate stem cells, which they can readily become. Experiments are yet to be performed to determine irrefutably whether they have the same, or a reduced, stem cell potential. These considerations inevitably also involve the question of whether stem cells differ intrinsically from transit cells. These observations with clonogenic stem cells could support the concept that they do not differ intrinsically but rather are instructed by their environment to act as stem cells: The cells higher up the lineage find a vacant stem cell niche or environment and, when in it, become stem cells. However, it would appear that there are definitely some transit cells that cannot function in this way, perhaps because they are too far from the niche.

However, there is now accumulating evidence for some very specific differences in stem cells that suggests they do normally divide asymmetrically and that they also selectively retain one copy of their DNA strands at cell division.

Differentiation options: pluripotency

Differentiation can be regarded as a qualitative change in a cell relative to others in the issue (i.e., it is a change in the appearance of the cell or its behaviour or function). Adult tissue stem cells are undifferentiated relative to most other cells in the tissue but may be differentiated relative to embryonic stem cells. Differentiation events generally occur at various points within the cell lineages derived from stem cells (i.e., either within the dividing transit population or later in the process of maturation). Thus, differentiation is a process not particularly associated with the stem cell compartment but with the progeny that are derived from the stem cells. This has led to some confusion because stem cells are often referred to as possessing a *totipotency* (embryonic stem cells) or a *pluripotency* (e.g., bone marrow stem cells), in terms of differentiation, but strictly it is not the stem cells that possess this ability but their progeny.

Clearly embryonic stem cells possess the ability to produce daughters that differentiate down many different pathways. Indeed, early embryonic cells are capable of making all the tissues of the body and if these cells are separated early in development, twins are formed that have completely normal differentiated and functional tissues. It is a matter of considerable interest and debate at present as to whether adult tissue stem cells have a limited and defined potential to produce daughters that differentiate down a limited range of pathways or whether they are totipotent in the same way that embryonic stem cells are. The instructive signals to induce specific differentiation pathways are largely unknown, which explains some of the difficulty in obtaining specific differentiation in a controlled way with embryonic stem cells grown in culture. However, the animal cloning experiments (such as those that created Dolly the sheep) clearly indicate that a relatively mature nucleus placed in the right environment (in this case in cytoplasm of an egg) can receive all the necessary instructive signals for differentiation to produce a viable, healthy organism. This process is still relatively 'hit and miss' in terms of success rate, with many nuclear transfers failing. However, it does indicate that the DNA in a relatively mature cell retains all the necessary instructions in the DNA to generate a normal organism and that this can be unmasked and activated, as long as the appropriate external instructive signals are provided.

It remains a matter of considerable debate whether any totipotential stem cells remain in the adult. Are such totipotential stem cells circulating throughout the body? Can they seed themselves in the various tissues of the body and produce the necessary differentiated cells for that tissue? Finally, are such cells extractable and can they be used to clinical advantage by grafting them back into patients or patient's tissues where serious deficiencies exist? It is clear from experiments in mice where bone marrow cells have been transplanted into animals where the endogenous bone marrow stem cells have all been killed by a dose of radiation, or in human patients who have received bone marrow grafts (transplants) from one sex to another that enable the transplanted stem cells and their progeny to be recognised by virtue of the presence or absence of the Y chromosome, that cells with the marker chromosome can appear in a wide variety of tissues and can express differentiation markers specific for that tissue. This has led to the term *stem cell plasticity* to refer to the ability of these bone-marrow-derived cells to produce unusual differentiated progeny. However, these experiments are subject to much criticism and debate. The marked bone-marrow-derived cells that appear in unusual tissues are generally isolated single cells that could express markers of novel differentiation (transdifferentiation). However, these isolated marked cells exhibit none of the proliferative characteristics of a stem cell (neither steady-state proliferation nor regenerative ability). However, in many of the experiments, fusion of a circulating bone-marrow-derived cell with some local cell, the cell death of a circulating cell with phagocytosis of its DNA fragments by a host cell, or even the specific transfer of chromosomes or chromosome fragments from one cell to another cannot be ruled out.

The best example of transdifferentiation of circulating bone marrow cells involves the repopulation of the liver following liver damage. It is possible in some circumstances for about 40 percent of the liver cells to express bone marrow stem cell-specific markers. These cells have all the attributes of fully functional hepatocytes. However, if the liver is subsequently redamaged and regeneration is forced, the regeneration occurs specifically from the non-bone-marrow-derived liver cells. Thus, these bone-marrow-derived liver cells lack the fundamental regenerative property attributable to most stem cells.

This is a difficult and rapidly developing field. It seems highly likely that with further developments it will be possible to provide the necessary environment and instructive signals that will enable adult tissue stem cells to make a wide diversity of tissues. If and when this is achieved, the distinction between adult tissue stem cells and embryonic stem cells becomes blurred. At present, bone marrow stem cells in an adult are the stem cells that process the greatest range of differentiation options for their progeny. It is clear that a common ancestral stem cell is capable of producing progeny that differentiate down all the haemopoietic cell lineages (erythrocytes, megakaryocytes, basophils, mast cells, eosinosylophils, neutrophils, and macrophages). In addition to these, there is evidence that bone marrow stem cells are the precursors for the lymphoid tissue, certain fibroblasts populations, and some components of the bone.

One of the properties that seem to be associated with stem cells is their ability to produce a variety of different differentiated progeny or lineages. This is most dramatically demonstrated for the haemopoietic stem cell that can form a whole range of different cell types and, of course, is even more dramatically expressed when one considers stem cells in the developing embryo. In the intestine at least four very distinct cell lineages originate from the stem cells that are located at a specific position in the crypt. These are the predominant columnar epithelial cells, the mucous secreting goblet cells, the Paneth cells, and a hormone-secreting lineage known as the enteroendocrine cells. There is also some indication that a specialised cell type known as M cells, found particularly in the epithelium covering collections of lymphocytes (lymph nodes) is also derived from the same stem cell compartment. This ability to produce multiple differentiated lineages is referred to as pluripotency for differentiation.

Further reading

Bjerknes, M., and Cheng, H. Modulation of specific intestinal epithelial progenitors by enteric neurons. *Proc. Natl. Acad. Sci. USA* **98**: 12497–12502, 2001.

Bjerknes, M., and Cheng, H. Clonal analysis of mouse intestinal epithelial progenitors. *Gastroenterology* **116**: 7–14, 1999.

Bjerknes, M., and Cheng, H. The stem cell zone of the small intestinal epithelium. *Am. J. Anat.* **160**: 51–63, 1981.

Cai, W., Roberts, S. A., Bowley, E., Hendry, J. H., and Potten, C. S. The differential survival of murine small and large intestinal crypts following ionising radiation. *Int. J. Rad. Biol.* **71**: 145–155, 1997.

Cai, W., Roberts, S. A., and Potten, C. S. The number of clonogenic cells in three regions of murine large intestinal crypts. *Int. J. Rad. Biol.* **71**: 573–579, 1997.

Cairnie, A. B., Lala, P. K., and Osmond, D. G. (eds.). Stem Cells of Renewing Cell Populations. Academic Press, New York, p. 389, 1976.

Cheng, H., and Leblond, C. P. Origin differentiation and renewal of the four main epithelial cell types in the mouse small intestine. Am. J. Anat. **141**: 531–561, 1974.

Gilbert, C. W., and Lajtha, L. G. The importance of cell population kinetics in determining response to irradiation of normal and malignant tissue. In: Cellular Radiation Biology, M. D. Anderson (ed.), Williams & Wilkins, Houston, pp. 474–495, 1965.

Hall, P. A., and Watt, F. M. Stem cells: the generation and maintenance of cellular diversity. *Development* **106**: 619–633, 1989.

Lajtha, L. G. Stem cell concepts. *Differentiation* **14**: 23–34, 1979.

Lord, B. I. Biology of the haemopoietic stem cell In: Stem Cells, C. S. Potten (ed.). Academic Press, London, pp. 401–422, 1997.

Marshman, E., Booth, C., and Potten, C. S. The intestinal epithelial stem cell (our favourite cell). Bioessays **24**: 91–98, 2002.

Mills, J. C., and Gordon, J. I. The intestinal stem cell niche: there grows the neighborhood. *Proc. Natl. Acad. Sci. USA* **98**: 12334–12336, 2001.

Potten, C. S., Owen, G., and Booth, D. Intestinal stem cells protect their genome by selective segregation of template DNA strands. *J. Cell Sci*. **115**: 2381–2388, 2002.

Potten, C. S., and Hendry, J. H. (eds). Stem Cells: Their Identification and Characterisation. Churchill-Livingstone, Edinburgh, p. 304, 1983.

Potten, C. S., and Hendry, J. H. The microcolony assay in mouse small intestine. In: Cell Clones, C. S. Potten and J. H. Hendry (eds.). Churchill-Livingstone, Edinburgh, pp. 155–159, 1985.

Potten, C. S., and Hendry, J. H. Clonal regeneration studies. In: Radiation and Gut, C. S. Potten and J. H. Hendry (eds.). Elsevier, Amsterdam, pp. 45–54, 1995.

Potten, C. S., and Loeffler, M. Stem cells: attributes, cycles, spirals, pitfalls and uncertainties: lessons for and from the crypt. *Development* **110**: 1001–1020, 1990.

Potten, C. S., Schofield, R., and Lajtha, L. G. A comparison of cell replacement in bone marrow, testes and three regions of surface epithelium. *Biochem. Biophys. Acta* **560**: 281–299, 1979.

Potten, C. S. Stem cells in gastrointestinal epithelium: numbers, characteristics and death. *Phil. Trans. Royal Soc. Lond.* (B) **353**: 821–30, 1998.

Potten, C. S. Structure, function and proliferative organisation of mammalian gut. In: Radiation and Gut, C. S. Potten and J. H. Hendry (eds.). Elsevier, Amsterdam. pp. 1–31, 1995.

Potten, C. S., Booth, C., and Pritchard, D. M. The intestinal epithelial stem cell: the mucosal governor. *Int J. Exp. Pathol.* **78**: 219–243, 1997.

Till, J. E., and McCulloch, E. A. A direct measurement of the radiation. sensitivity of various mouse bone marrow cells. *Radiat. Res.* **14**: 213–222, 1961.

Withers, H. R., and Elkind, M. M. Microcolony survival assay for cells of mouse intestinal mucosa exposed to radiation. *Int. J. Rad. Biol.* **17**: 261–267, 1970.

Wong, M. H., Rubinfield, B., and Gordon, J. I. Effects of forced expression of an NH_2-terminal truncated β-catenin on mouse intestinal epithelial homeostasis. *J. Cell Biol.* **141**: 765–777, 1998.

Wright, N. A. Aspects of the biology of regeneration and repair in the human gastrointestinal tract. *Phil. Trans. Royal Soc. Lond.* (B) **353**: 925–933, 1998.

7

An *in vivo* system to study apoptosis: the small intestine

Proliferative organisation in the gut

We have seen how proliferative cells pass through the sequential process of the cell cycle and how there may be a hierarchy of functional potential amongst the proliferative cells in a tissue. There may be only a small number of lineage or hierarchy ancestor stem cells within a tissue (see Chapter 6). The gastrointestinal tract, particularly the small intestine, is a tissue in which these cellular properties have been extensively studied and the hierarchies are, therefore, fairly well understood. Consequently, it is a tissue that is a useful model with which to study the characteristics of stem cells, their response to injury, their regenerative capacity, and their location and numbers. It is also a useful system for studying the functional role of apoptosis within an integrated, multicellular system. It is a tissue with which I have worked for a large number of years and for this reason I should like to describe it and summarise some of the experiments that have been performed in this tissue in relation to apoptosis.

The gastrointestinal tract is effectively a tube that runs from the mouth to the anus. I restrict my comments to the small and large intestine, which are separated at the point where the appendix occurs. The small intestine is by far the largest region of the gastrointestinal tract and contains the most rapidly proliferating cells. In the mouse, the small intestine is about 20 cm in length, whereas in humans it is about 6 m. This tube consists of an epithelial mucosa sitting on a connective tissue base surrounded by various muscle layers. The epithelium or mucosa is a simple columnar epithelium, one cell thick. Columnar epithelial cells have a cylindrical shape, with an apical membrane greatly extended in surface area by being folded into microvilli, the so-called brush-border. The microvilli are rather like the pile on a carpet; they are densely packed fingerlike projections of the cell that greatly increase the absorptive surface area. The individual columnar cells are thus polarised, the outer surface differing dramatically from the inner surface, which is attached to a basement membrane that consists of specialised extracellular proteins, generated by both the epithelial cell and the underlying fibroblasts in the connective tissue. The epithelium is folded in a complex fashion generating flask-shaped bags of cells called 'crypts', which extend down into the connective tissue. In the colon, the crypts open onto the flat table region, providing a relatively smooth surface to the inside of the colon. In the small intestine, the crypts extend into large fingerlike projections called villi that protrude into the lumen (the centre of the tube) of the small intestine. Thus, the epithelium as a whole is also highly polarised with an outer region, the villi, exposed to the contents of the intestine and an inner region, the crypts, embedded deep in the supportive connective tissue. This structural or architectural configuration is illustrated in figure 32.

Cell proliferation is restricted to cells in the mid and lower regions of the crypt. In the mouse, small intestine cells can be demonstrated to be passing through the cell cycle twice a day; that is, their cell cycle time (duration) is twelve hours: this is relatively fast compared to other types of cell. Approximately half of the time it takes to complete the cell cycle is taken up with replicating the DNA, the synthesis or S phase. Thus, if thymidine or bromodeoxyuridine incorporation is used to identify the S-phase cells (see Chapter 2), approximately half of the cells this midregion of the crypt will be positively labelled. Fewer studies have been undertaken to measure cell cycle time in the mouse large intestine, but it would appear to be two or three times slower, at between thirty-six to forty-eight hours. In the human, these times are between four to eight times slower still, although data are limited.

Throughout the gastrointestinal tract, cells differentiate to form various functional cell types. In the small intestine, the major

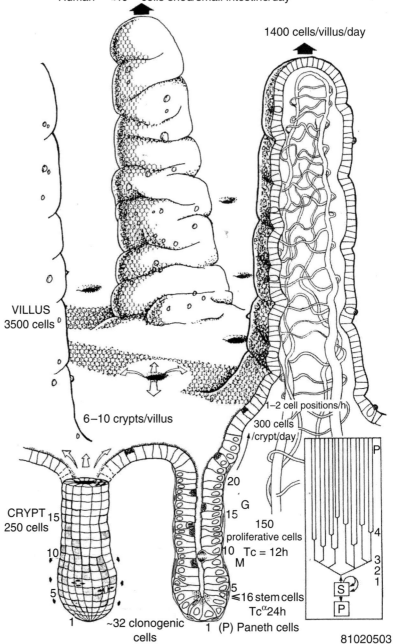

32. Cellular and structural architecture of the small intestine, showing the proliferative units, crypts, and the functional projections, the villi. A cross-sectional view shows the simple columnar epithelium and the intricate capillary network. The cell lineage organisation within the crypt is illustrated with the lineage ancestor stem cells. The position of a cell in the lineage can be directly related to its position in the tissue. Reproduced with permission and slight modifications from Potten, 1995.

functional cell population are columnar epithelial cells whose role is nutrient absorption. In both the small and large intestine, numerous mucous-secreting cells are also formed: mucous provides both lubrication for movement of intestinal content and protection against possible colonisation of the intestinal surface by pathogenic microorganisms. At the base of the small intestinal crypts in mice and humans, a third differentiated cell type called Paneth cells are found. Paneth cells play a role in innate immunity, producing antimicrobial peptides and enzymes, which are stored in granules within the cells that are easily visible using light microscopy. Paneth cells appear to be lacking in certain species (e.g., dogs and pigs) and their role may be fulfilled by other cells. There are other differentiated cell populations, although these are less numerous. The crypts are, as far as we know, closed units or organs with no cellular input from outside the structure, once they are formed during development. They must, therefore, contain self-maintaining stem cells that can sustain the structure for essentially the life of the animal. The crypts in the small intestine of mice are small, containing a total of no more than 250 cells. In the large intestine they are larger, the number of cells depending somewhat on the site within the large intestine but in the mouse the crypts in the midcolon contain between 350 and 400 cells. Again in the human, the structures are larger by factors of between 2 and 4. In the mouse small intestinal crypt, 150–160 of the 250 or so cells are passing through the cell cycle at any one time. Additionally, there are about 30–40 Paneth cells at the base of the crypt, including a few undifferentiated cells scattered amongst the Paneth cells that are probably early Paneth precursors, that is, cells in the Paneth lineage. In the upper regions of the crypts there are about another 50 nonproliferating, maturing columnar cells and goblet (mucous-secreting) cells that are in the process of migrating onto the villus.

Studies where proliferating cells in the crypt are labelled, with, for example, tritiated thymidine and samples are taken not immediately but after varying periods of time, illustrate two important points about this system. Firstly, that the cells are migrating or moving continuously from the crypt onto the villus. These labelling studies can be used to measure the velocity of the cell movement, which is approximately one cell diameter per hour at the top of the crypt. Secondly, these studies show that the cells in the upper region of the crypt only divide once after incorporating the tritiated thymidine before they move onto the villus (to become differentiated cells); the cells a little lower in the crypt divide twice and the cells below them three times. This is evidenced by the dilution of the incorporated radioactive label within the cells, as they move from the crypt to the villus (i.e., along the crypt–villus axis). So, these studies show that the cells are moving inexorably from the proliferative compartment to the differentiated compartment. Consequently, cells in the middle and upper regions of the crypt have a very restrictive division potential and therefore would be poor candidates to be characterised as stem cells. Another interpretation from these observations is the conclusion that cells born in the mid and upper regions of crypt have a very limited life expectancy. The data in the mouse small intestine suggest that these cells reach the tip of the villus, from where they are extruded into the lumen, about three days after being born in the crypt. They don't live very long!

So, if the cells in the upper region of the crypt have a limited potential and are not candidates for stem cells, where are the self-maintaining stem cells, and how many of them are there? The same studies where cells are labelled and then their fate with the passage of time is determined can be used to address the following question: 'Where is the origin of all the movement and migration; that is, where in the crypt does everything stem from?' The answer, which is illustrated in figure 33, shows that the origin appears to be at about the fourth position from the base of the crypt in the small intestine, which is immediately above the Paneth population. The position of the back extrapolate on the cell migration velocity studies as shown in figure 33, which is the point of origin of all the movement, varies somewhat from experiment to experiment. In the data shown in figure 33, for the small intestine, the line extrapolates back to close to the fifth cell position; in other cases the value might be three or four. The general conclusion from these experiments is that the origin is at about the fourth position from the base of the crypt.

AN *IN VIVO* SYSTEM TO STUDY APOPTOSIS

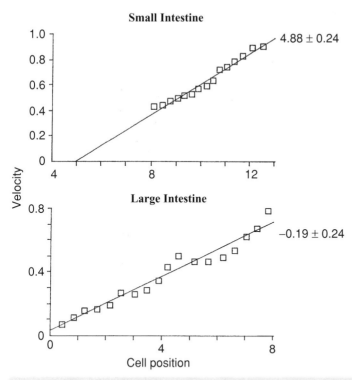

33. Experiments have been performed to determine the cell migration velocity in both the small and large intestine. Frequency plots for the cell position of the tritiated thymidine-labelling index were determined at different times after labelling, from which the velocity at each cell position was determined. The origin of the velocity versus cell position plot indicates the origin of migration, cell production, and the position of the stem cells. Modified from Potten *et al.*, 1997.

In the large intestine, the origin is at the very base of the crypt, at least in the midcolonic region of the mouse. From here, cells migrate and divide a certain number of times as they move up the crypt, until they reach the upper third of the crypt, where they cease division activity and start to differentiate into mature functional cells. They then continue their migration onto the villus and "march-on" inexorably towards the tip, from which they are extruded like the apocryphal lemmings, leaping off the edge of a cliff. Once in the lumen of the gut they are digested and their cellular components, including the DNA bases, are very efficiently recycled.

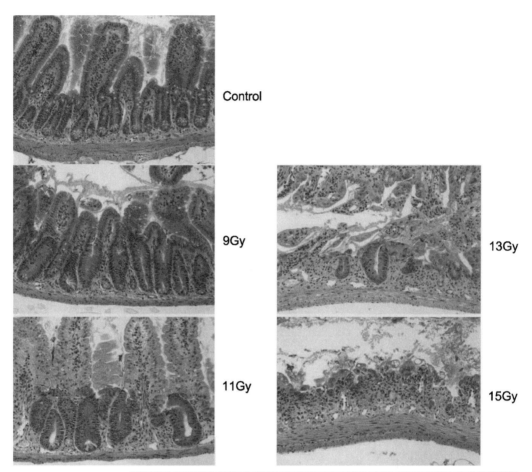

34. Haematoxylin-and-eosin-stained sections of mouse small intestine (control and four days after various doses of radiation). In the control, the crypts and the villi can be clearly seen. In the postirradiated samples, the regenerating crypts can be seen. They decrease in number as the dose is increased. They are larger in size at day four than the control crypts.

The ability of some cells to regenerate the crypt *via* a process of clonal regeneration, which was described in chapter 6, is a test of the functional potential of stem cells. This has been studied extensively in the small intestine of the mouse. Regenerating clones can be seen easily in histological sections of the small intestine, three to four days after exposure to cytotoxic doses of radiation or drugs and five to six days after exposure to radiation in the midcolonic region of the large intestine (see figure 34). The number of these colonies in a

AN *IN VIVO* SYSTEM TO STUDY APOPTOSIS

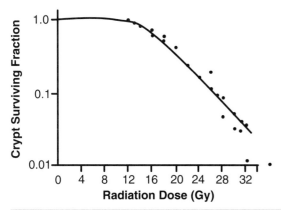

35. A radiation dose–response curve for surviving regenerating crypts in the small intestine.

unit of area of intestine, or more easily in a unit of length, can be determined. A convenient unit of length is the outer circumference of a transverse section of the intestine. The dose of radiation can be varied and a dose–response curve, or a survival curve, for the regenerating crypts and for the clonogenic cells can be determined. These survival curves, as shown in figure 35, have a characteristic broad shoulder, where the number of surviving crypts or regenerating clones remains unchanged until a certain dose has been received (usually about 9 Gy), after which the number of surviving clones decreases exponentially with increasing radiation dose, as the last surviving clonogenic cell per crypt is killed (sterilised). The slope of this exponential region of the curve is a measure of the radiation sensitivity of the cells. The reciprocal of the slope, called the D_0, is commonly used to define these curves together with the extrapolate back to zero dose, which is usually designated as N, the extrapolation number.

The shoulder, where there is little change with increasing dose, is attributable to the fact that each crypt contains more than one cell capable of undergoing regeneration. In fact crypts contain a number of such clonogenic stem cells, and each of these has the ability to repair a certain amount of radiation damage before it survives and regenerates the crypt or is reproductively sterilised or killed. Crypts do not disappear until the last of these regenerative clonogenic cells is killed. The point where the dose–response curve changes shape (i.e., the edge

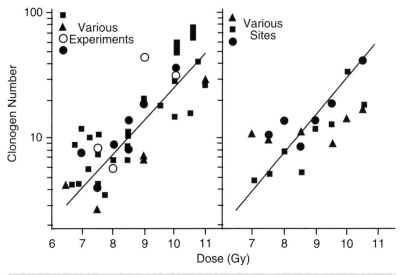

36. Estimates for the number of clonogenic stem cells per crypt in the small and large intestine showing the variation in the estimates with the dose of radiation used to determine the number (i.e., variation with the severity of the damage induced). Modified from Potten et al., 1997, and Cai et al., 1997.

of the shoulder), is the point at which, on average each crypt contains only one more surviving cell, all the others having been killed. Now, the survival of crypts follows the pattern of survival of these last regenerative clonogenic stem cell. Data can be generated, which after various limited assumptions are made, can provide information on the number of clonogenic cells in each unirradiated crypt. The data that are obtained are surprising (figure 36) because the answer to the question of how many clonogenic cells are there per crypt appears to depend on the dose of radiation that is delivered to determine the number. In other words the number of cells capable of regenerating the crypt depends on the severity of the damage that is induced. The greater the damage, in terms of cell kill, the more cells capable of regenerating the structure are found. The data suggest that following exposure to moderate doses of radiation, the crypt contains about six cells capable of regeneration. If the levels of radiation (i.e., DNA damage) are increased, a greater number of cells appear to be capable of regenerating the crypt, up to as many as thirty to forty. Higher levels of damage do not seem to be capable of recruiting any further cells.

AN IN VIVO SYSTEM TO STUDY APOPTOSIS

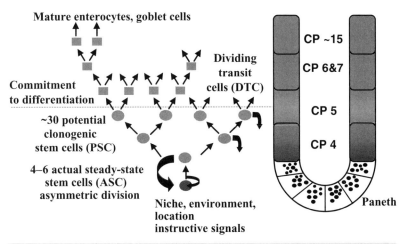

37. Crypt stem cell hierarchy. Diagram illustrating the cell lineages believed to operate in the proliferative units of the intestine, the crypts. In the small intestine of the mouse, between four and six lineages are believed to be packed into each crypt, therefore, there are four to six lineage ancestor stem cells that exist under steady-state conditions. However, the commitment to differentiation does not occur until the third generation in the lineage, which means there is a greater number of potential stem cells. The position of a cell in the lineage can be related to its position in the crypt, with the actual and potential stem cells occupying the fourth and fifth positions from the crypt base.

Mathematical modelling of the sort of data illustrated in figure 33 for cell migration, and the modelling of many other cell proliferation experiments, together with studies on the effects of irradiation, have resulted in the conclusion that the crypt could function perfectly adequately with a small number of self-maintaining, lineage-ancestor cells or stem cells. Their actual numbers are somewhat uncertain, but probably there are about four to six per crypt. The current view is that the crypt is organised as follows; the cells within the proliferative compartment are organised into a series of cell lineages of the sort illustrated in figure 37. It is believed that there are between four and six such lineages in each small intestinal crypt in the mouse; that is, there are between four and six lineage ancestor stem cells, which are responsible for all day-to-day cell production. They divide essentially asymmetrically, generating a daughter cell that enters the dividing transit population and another daughter, which is responsible for self-maintenance of the stem cell population (i.e. they remain as stem cells). They may occasionally divide symmetrically, producing an extra stem cell. Under normal circumstances this would need to be removed by either apoptosis or differentiation; we return to this subject in this chapter. If all six of these ultimate stem cells are destroyed by a cytotoxic exposure (e.g., ionising radiation), some of the cells in the early part of the dividing transit population are sufficiently undifferentiated to be able to revert and assume stem cell function. If all the early transit cells are also killed, there is a further population of up to about twenty-four cells in the early transit lineage that can also revert. The hierarchical organisation thus provides considerable flexibility. It is, however, unlikely that under normal circumstances these third-tier stem cells are called into action. This is only seen in laboratory conditions. The ultimate stem cells, responsible for day-to-day work in terms of cell production, are the cells with the greatest sensitivity to damage induced by radiation and easily undergo apoptosis. Their immediate daughters that can replace them have a greater resistance or better repair capacity following radiation exposure. The third tier of cells that can replace these cells, which constitute the second and part of the third transit generation, are even more resistant to damage. This structure is illustrated in figure 37. This represents the current model or hypothesis for how the stem cells are organised in the small intestinal crypt of the mouse. It should be realised that situations like those when we irradiate mice to perform clonal regeneration studies, do not occur in nature. Under natural conditions the crypt is probably successfully maintained for most of the life of the animal by the four to six ultimate stem cells.

A variety of cell proliferation studies have clearly shown that the cells at the origin of the cell migration pathways have a slower cell cycle than most of the crypt cells. The different experiments tend to generate slightly different values for the stem cell cycle time but it is probably about twenty-four hours, or within the range sixteen to thirty-six hours.

As mentioned earlier, the stem cell population has no specific markers or identifying features but one of the attractive characteristics of the intestinal mucosa as a model system is that one can relate the position of a cell in the tissue to its hierarchical status. This is illustrated in figure 32 and in more detail in figure 37, which represents a

AN *IN VIVO* SYSTEM TO STUDY APOPTOSIS

schematic diagram of a longitudinal section through a crypt. If such sections through crypts are selected and analysed, the characteristics of the cells at each position (as shown by the numbering) can be recorded. As illustrated in this diagram, any particular feature that one is interested in can be recorded on a cell positional basis; this could be cell proliferation markers, differentiation markers, or apoptosis. The diagram shows the Paneth cells at the bottom of the crypt, the goblet cells differentiating in the middle and upper regions of the crypt, cells undergoing mitosis, and apoptotic bodies. The graph shows two typical sets of data, one illustrating the distribution of a proliferation marker, for example, the fraction of cells labelled by tritiated thymidine or the fraction of cells in S phase, and the occurrence of apoptotic cell death following a small dose of radiation. The data can be generated as a frequency plot and the activity or characteristics of cells at a particular position can be analysed. The data show how the bulk of proliferating cells are distributed in the midregions of the crypt between cell positions five and fifteen, and that the apoptosis generated by a small dose of radiation has its maximum effects over cell positions three to five, but we will return to this in a later chapter. A frequency plot can be generated for the Paneth cells. This shows that the Paneth cells have a fairly scattered distribution and in one crypt section may cease to appear at cell position two or three, whereas in another crypt section they may continue to be present until about cell position seven. It is believed that the stem cells are located in the first non-Paneth cell position in the crypt, with the four to six ultimate stem cells scattered in an undulating annulus or ring that occurs in this first non-Paneth position, which on average is cell position four. This annulus may contain up to sixteen cells (the number of cells in a crypt circumference), including the four to six ultimate stem cells. Mathematical modelling exercises can generate the theoretical spatial distribution of the ultimate stem cells and the potential clonogenic stem cells and this is illustrated in figure 38. Figure 39 is a schematic representation of the three dimensional architecture at the base of the crypt, showing the Paneth cells, and how they vary their position as one goes round the circumferential annulus and how the ultimate stem cells may appear at the second position or the sixth position,

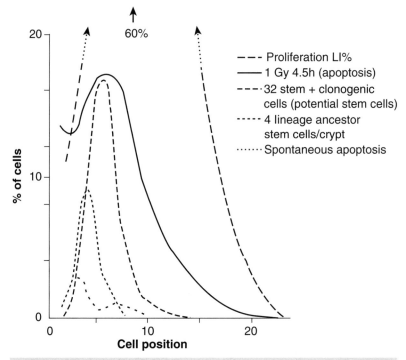

38. The theoretical distribution of stem cells in the small intestinal crypt of the mouse based on mathematical modelling studies. The graph shows the frequency distribution (i.e., the spread in cell positions) for the actual lineage ancestor stem cells and the potential clonogenic stem cells (from mathematical modelling studies with Dr. U. Paulus and Professor M. Loeffler, Cologne and Leipzig). Also shown are the cell positional frequency plots for proliferative cells (S-phase labelling index; LI), spontaneous apoptosis, and the apoptosis induced by a dose of 1 Gy of radiation.

but the average will be the fourth. The diagram also illustrates the hypothetical distribution of clonogenic stem cells within the dividing transit population, specifically within generations one, two, and three of the dividing transit population. This diagram illustrates the complexity of the tissue architecture and the complex cellular interactions and communications that must exist to ensure that the structure is rigidly maintained. The crypts all tend to have a similar size and shape and so there cannot be wild fluctuations in the number of stem cells and the number of lineages derived from the stem cells. One of the major challenges for the next decade is to try and understand the intercellular communications and chemical signals that ensure that each crypt maintains only four to six ultimate stem cells, produces

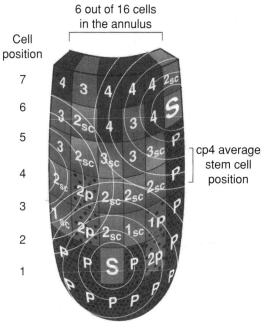

P = Paneth cell
1p, 2p = Paneth lineage cells
S = Stem cell
1_{sc} = 1st transit lineage clonogenic stem cell
2_{sc} = 2nd transit lineage clonogenic stem cell

39. Diagram of the small intestinal crypt showing the spatial distribution of the lineage ancestor stem cells. The numbers of these crucial cells per crypt must be tightly regulated or the crypts would vary enormously in size, because each stem cell generates a lineage of up to 128 daughter cells. The regulation in numbers may be determined by the levels of a stem cell signal, such as illustrated by the concentric rings on the diagram. Reproduced with permission from Marshman et al., 2002.

the right number of Paneth cells, goblet cells, and columnar cells, and rarely develops cancer.

There is some evidence that crypts may die and are gradually replaced throughout life by the splitting of preexisting crypts (crypt fission or budding). However, this process is rare and may account for only the turnover or replacement of a few crypts in the life of a mouse.

There are some who postulate that the true crypt stem cells are located scattered amongst the Paneth cells (the undifferentiated intercalated cells). However, in our view, these are more likely to be cells in the Paneth cell lineage and that the true location of the stem cells is immediately above the Paneth cells at cell position four. The supporting observations for this are as follows:

1. Lineage tracking studies with tritiated thymidine suggest that the origin of all movement is cell position four.
2. The hypothesis that the stem cells are at cell position four involves less complex cell movement processes. If the stem cells were within the Paneth compartment, their daughters would have to move up between the Paneth cells to reach a position from which they could generate the rest of the epithelial cells, whereas any Paneth precursors would have to move down (i.e. in the opposite direction).
3. Regeneration of the crypt following various cytotoxic insults can be seen to originate from cell position four. This is particularly clear twenty-four hours after high doses of a drug called 5-fluorouracil.
4. Label-retaining cells and cells that segregate their DNA in a special way hypothesised for stem cells occur at cell position four.
5. Some stem and early lineage markers, such as musashi-1, are expressed at cell position four.
6. There is a very specific and highly sensitive apoptosis-susceptible cell population at cell position four. It has been suggested that these are cells operating a stem cell genome protection mechanism.

The regulation of differentiation and ordered cell movement in the crypt is now beginning to be understood and these processes seem to be governed by the interaction of proteins such as the T-cell factor (TCF) family, β-catenin, c-myc and p21, and ephrin ligands and their receptors (EphB2 and EphB3). β-Catenin has two roles in the cell. It can combine with E-cadherin and promote cell-to-cell adhesion. Alternatively, it can combine with TCF proteins, bind to DNA, and increase the expression of genes such as those coding for c-myc in order to drive proliferation. The levels of β-catenin decline in cells further up the crypt so that proliferation is switched off and differentiation

AN *IN VIVO* SYSTEM TO STUDY APOPTOSIS

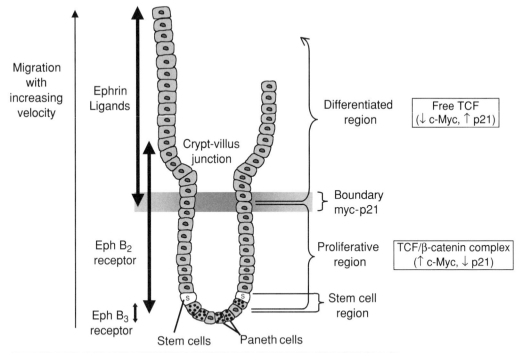

40. Diagram showing the relationships among the expression of β-catenin/T-cell factor (TCF) proteins and ephrins and ephrin receptors, which govern epithelial cell migration and differentiation within the small intestinal crypts.

is permitted to occur. The proliferating cells that express the higher levels of β-catenin/TCF express ephrin receptors and the differentiated cells with their reduced β-catenin/TCF levels express ephrin ligands, which are secreted. It is the interaction of the ephrin ligands and their receptors that somehow regulates the ordered migration of cells within the crypts (i.e. tells them what direction to travel in and how fast). These interactions are shown in figure 40.

Apoptosis in the gut

Spontaneous and radiation-induced apoptosis

We have seen how the gastrointestinal system is a multicellular tissue that can be employed as a useful cell biological model to study many processes and regulatory mechanisms, including apoptosis. What I want to do in this chapter is summarise the results of many years' study of apoptosis in the gut, and the genes involved in this process, under a variety of conditions. During the course of this chapter, I attempt to answer the following questions: Which cells in the tissue are dying? Why are they dying? Does the tissue recognise that they are dead? And how does the tissue respond to this information? In other words, what is the biological significance or role of apoptosis within the framework of a tissue like the gut?

First of all, in the small intestine of a mouse or a human, a few cells in the crypts are undergoing apoptosis at any moment in time, even though this is in a situation where, as far as we know, no exposure to a damaging agent (e.g., ionising radiation) has occurred (figure 41). The fact that this *spontaneous apoptosis* is independent of DNA damage is further supported by the observation that in animals where the *p53* gene (which is involved in DNA-damage response; see Chapter 5) has been transgenically deleted have the same levels of background apoptosis in their intestinal crypts. This background or spontaneous apoptosis in the crypts involves the deletion of perfectly healthy intestinal epithelial cells. As discussed previously, large numbers of perfectly healthy cells are deleted by apoptosis during the later stages of embryonic development, which is essential for tissue restructuring and reorganisation. This type of developmental apoptosis also occurs normally in the p53 knockout animals. We suspect that because both these situations, development and adult steady state, involve the apoptosis of healthy cells, apoptosis occurring in these two contexts may be regulated in a similar fashion.

If the number of spontaneous apoptotic cells or fragments in the intestine are counted and expressed as a percentage of all the crypt cells, the incidence of this type of cell death is very rare (less than 1 percent of the crypt cells, which is equivalent to one apoptotic event in perhaps every fifth intestinal crypt section in the mouse). Figure 41 shows the appearance of these apoptotic bodies, each representing the death of a cell in mouse and human crypt sections. If we now ask the question, 'Where in the crypt, at which cell position, does this spontaneous apoptosis most frequently occur?', the answer is difficult to obtain. This is because spontaneous apoptosis has a very low incidence; however, by pooling data from many experiments the

AN IN VIVO SYSTEM TO STUDY APOPTOSIS

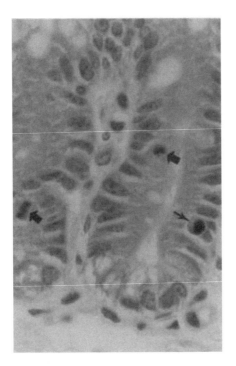

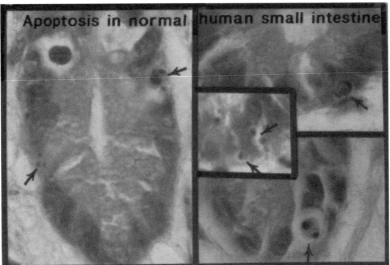

41. (Top panel) Haematoxylin-and-eosin-stained section of a normal, healthy mouse showing cells in mitosis (large arrows) and a single apoptotic cell (small arrow). (Bottom panel) Similar spontaneous apoptosis can also be observed in normal, healthy human small intestine.

conclusion can be drawn that it occurs with the greatest frequency at about cell position four, where the stem cells are believed to be located. Because each section cut longitudinally through the centre of the flask-shaped crypt only contains two cells at cells position four (one on either side of the crypt) the apoptotic counts, expressed relative to the number of cell position four cells, show quite a significant rate of cell death. Observations suggest that between 1 and 10 percentage of the cells at cell position four (the stem cells) are undergoing apoptosis without exposure to any damaging agent at any one time. An actual cell position distribution for this spontaneous apoptosis is shown in figure 38.

In the large intestine of the mouse, spontaneous apoptosis is *very* rare. The frequency is about ten times lower than in the small intestine and those apoptotic cells that are observed tend to be randomly distributed throughout the entire crypt. In the midcolon region of the large bowel of the mouse the stem cells are believed to be located at the very base of the crypt (cell positions 1 and 2). Bcl-2, the antiapoptotic, cell survival-associated protein (see Chapter 5) is expressed by the cells at the very base of the crypt (the stem cell region: see earlier) in the colon of the mouse and humans. In humans it seems to be a consistent observation, in the mouse the expression of Bcl-2 in the crypt is more difficult to demonstrate reproducibly, with sometimes expression being observed and sometimes not. We believe that the expression of Bcl-2 by cells in the stem cell region suppresses the spontaneous apoptosis in the colon. The reason for the more variable expression of Bcl-2 in the mouse is unclear. Its expression may be being influenced by cyclic or variable interactions with other members of the Bcl-2 gene family.

Other sites where apoptosis is seen in healthy intestine are at the villus tip and in the table region between the crypts in the large intestine. Apoptosis of cells in these regions is related to the terminal stages in the epithelial cell's life. It appears that as cells approach the end of their functional (i.e., useful) life they switch on the apoptosis programme prior to, or coincident with, their extrusion from the epithelium into the lumen of the gut (see figure 11). The cells at the villus tip express some of the markers of apoptosis; for example, they activate the death genes *bax* and/or *bad*, and they stain positively for fragmented DNA when the TUNEL technique is used (figure 11). This phenomenon may also represent the natural consequence of the idea that the 'fall back' or 'default' state for all cells is apoptosis. As cells reach the end of their useful life, the survival signals that kept them alive may be removed or diluted and thus apoptosis is initiated.

In the small intestine, where Bcl-2 is not expressed, our current hypothesis is that the spontaneous apoptosis is responsible for removing occasional additional stem cells that may be produced excess to requirement of the crypt, for example, by an occasional symmetric division when asymmetric divisions is the norm. In other words the apoptosis is part of the regulatory process operating on stem cell numbers. Precisely how stem cell numbers are 'counted' is not defined but if there are too many, one or more receive the signal to commit suicide. The addition of a single extra stem cell in a crypt of 250 cells in the mouse may not seem an important phenomenon; however, each additional stem cell will inevitably produce between 64 and 128 additional cells in the lineage that it generates, and this would result in a major distortion in the crypt size and configuration. Mathematical modelling of the crypt suggests that occasionally (for example, 5 percent of the time) the stem cells may divide symmetrically. Thus, the spontaneous apoptosis ensures a stable crypt stem cell population and a stable crypt size. By implication, in the large intestine this regulatory process is not as rigidly controlled probably because of the expression of Bcl-2. As a consequence, in the large intestine stem cell numbers may gradually drift to higher values in any one crypt with the passage of time. Each time this happens a crypt would becomes larger, containing more cells. Such enlarged crypts are sometimes called hyperplastic, and the presence of hyperplastic crypts is thought by some to indicate either an early stage in the cancer development sequence or at least a state of increased cancer risk. Any increase in the number of stem cells would also mean that their would be more cells at risk of carcinogenic change (i.e., there would be more carcinogen target cells). An argument can be put that cancer is predominantly a disease of stem cells.

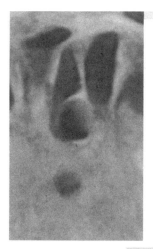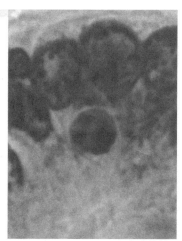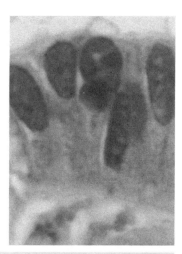

42. Three examples of apoptosis near the base of the small intestinal crypt (haematoxylin and eosin staining),

Radiation is a useful experimental cytotoxic agent, as the dose can be very carefully and precisely controlled. In 1977, I reported the observation that very small doses of radiation result in an elevation of the number of cells undergoing apoptosis in the crypts. Apoptosis was evident within a period of ninety minutes to two hours, with a peak incidence between three and six hours and that these apoptotic cells tended to occur in the lower regions of the crypts (figure 42). These observations have been subsequently repeated many times and in fact the peak incidence of apoptosis occurring at three to six hours is clearly associated with cell position four. This is true for a variety of sources of radiation delivered at difference dose rates: gamma rays and X-rays, which cause sparsely spaced ionisations in the nucleus, neutrons, which tend to cause a dense track of ionisations in the nucleus, and the very weak beta particles (emitted by tritiated thymidine incorporated into the DNA which irradiates essentially only the DNA) all show the same pattern. Certain cytotoxic drugs and some chemical mutagens or carcinogens also raise the apoptotic level with a cell position frequency distribution centred at the fourth cell position.

These radiation studies on the induction of apoptosis arose firstly from some initial observations, which indicated that incorporated

AN *IN VIVO* SYSTEM TO STUDY APOPTOSIS

tritiated thymidine could kill some cells in the crypt and, secondly, a desire to relate cell death to its inverse parameter, cell survival. The latter is generally measured by the clonogenic regeneration technique (i.e., by looking at the stem cells that survive) (see Chapter 6). It was quickly realised that there was no obvious relationship between the incidence of cell death observed at three to six hours after various radiation doses and the number of clonogenic stem cells that survive the radiation to repopulate and regenerate a crypt. However, by investigating progressively lower and lower doses, a dose–response relationship was observed for the number of apoptotic events at three to six hours. The results were surprising for a number of reasons. Firstly, they demonstrated an incredible radiosensitivity for some of the cells, doses as small as 1–5 cGy (0.01–0.05 Gy) raised the apoptotic levels dramatically. Secondly, the yield increased progressively with increasing dose, up to a value of about 50–100 cGy (0.5–1.0 Gy) when it appeared to saturate. Figure 38 shows a typical set of data for both the spontaneous apoptosis and the radiation-induced apoptosis, expressed on the basis of cell position, together with the distribution for the proliferative cells measured by the tritiated thymidine-labelling index. Also shown is the theoretical distribution for the ultimate stem cells and the clonogenic cell population, based on mathematical modelling studies. All the distributions are centred on cells around the fourth position from the bottom of the crypt. They have different heights and different spreads but the maximum value is, in each case, at around cell position four or five.

Radiation damage-induced apoptosis is initiated rapidly such that peak levels are observed within three to six hours of the radiation exposure (figure 43). After low doses, the natural background, spontaneous apoptosis levels are reestablished after twenty-four hours. If one addresses the question, 'Where in the crypt does this radiation induced apoptosis occur?', which is another way of asking the question, 'Which cells in the cell lineage die?', the answer is shown in figure 44, where it can be seen that the peak yield of apoptosis occurs in the small intestine at cell positions four to five. So, it is not the rapidly proliferating cells (shown by the dotted line in figure 44) but a subcompartment of the proliferative cells, specifically located

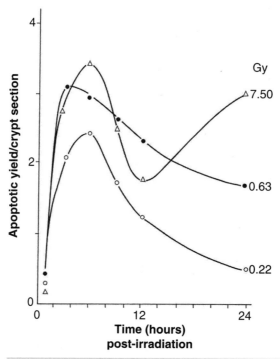

43. The relationship between apoptotic yield in the small intestinal crypt and time after exposure to small doses of radiation.

at positions four to five. This spatial distribution is quite similar to the theoretical distribution for the total stem cell compartment (see figure 38). Finally, one can address the question, 'How does the yield of apoptosis at four and a half hours after irradiation vary with radiation dose?'. The results were surprising and are shown in figure 45. Extremely small doses of radiation significantly raise the levels of apoptosis above the background spontaneous level. The yield increases with increasing dose, up to a value of about four apoptotic cells per crypt section (about six per whole crypt), at a dose of about 1 Gy, which is still a small dose when one considers the dose–response curve for the survival of clonogenic cells (see figure 35). If the data in the upper panel of figure 45 are expressed as a survival curve for the apoptosis-susceptible cells, we see the data as illustrated in the lower panel of figure 45. These data are unusual for mammalian cells, in terms of the high sensitivity and the simple, exponential relationship between survival and dose. This exponential relationship indicates

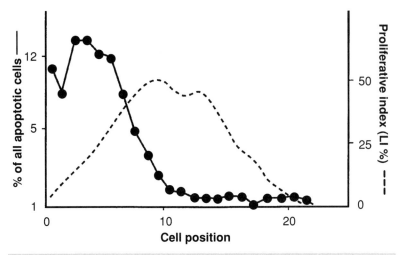

44. Cell positional distribution of apoptosis induced by small dose radiation.

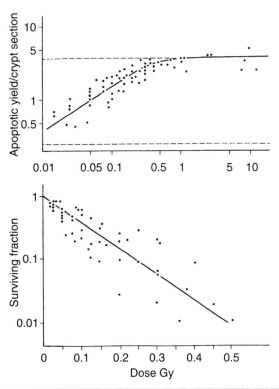

45. (Upper panel) Dose–response relationship for the apoptotic yield in the small intestine on a double logarithmic plot. The data points are taken from irradiation experiments covering a wide range of dose rates and radiation types. (Lower panel) The same data expressed as a survival curve for the apoptotic-susceptible cells, taken from the plateau at high doses in the upper panel.

that these cells do not have an ability to repair DNA damage, a conclusion born out by the fact the dose–response relationship shown in the upper panel shows no effect of variation in dose rate or the linear energy deposition of the radiation. It should be noted that this radiation-induced apoptosis (within the first six to eight hours following radiation exposure) is totally dependent on the presence of an active *p53* gene. If these experiments are done in mice where the *p53* gene has been deleted, no early, radiation-induced apoptosis is observed. In summary, these observations strongly suggest that the crypt contains a population of exquisitely sensitive cells at the stem cell position, in numbers equivalent to the estimated number of ultimate stem cells. These cells do not attempt repair but commit an altruistic suicide.

One further surprising feature of this system is that if a very small dose of radiation is used to kill a single cell at cell position four in the crypt, one can observe an almost instantaneous increase in the proliferation (S phase cells or mitotic cells) in the lower regions of the crypt. This suggests a very tight regulatory process such that small changes in stem cell numbers trigger a rapid compensatory reaction. It is almost if they were somehow counting their numbers and knew that one of their party was missing!

p53 dependence

There are several implications that can be derived from these experimental results. Firstly, the cells that die via apoptosis are extremely sensitive, which implies that they must have very efficient damage detection mechanisms. They can apparently detect a single damage event resulting from the interaction between an ionising radiation particle and the cellular DNA. The cells do not attempt to repair this damage but rather they commit suicide via apoptosis. This is an extremely efficient protection mechanism, because it eliminates the damage, as well as any chance of the cell making any mistakes during repair. This process is operating extremely efficiently in the small intestine and this may be one reason why there is such a low incidence of small intestinal cancer. The *p53* gene appears to be an essential element in this damage recognition–damage response process, because in animals where the *p53* gene has been deleted (the *p53* knockout or null mice) exposure to a dose of radiation generates no enhancement of the apoptotic yield over and above the spontaneous levels. Thus, the early apoptosis that is seen following radiation is totally p53-dependent. Apoptotic cells are also observed at later times (for example, twenty-four hours, particularly after higher doses of radiation), but this apoptosis appears to be *p53*-independent and is probably associated with the burst of proliferative activity associated with crypt regeneration (e.g., it involves the death of damaged clonogenic cells that attempt to divide as part of the stimulated regenerative proliferation) or it is a mechanism for deleting stem cells that are excess to requirement.

The wild-type p53 protein is rapidly synthesised in irradiated crypts; it appears with a time course and a cell positional distribution virtually identical to that for apoptosis. Peak levels are seen two to four hours after irradiation and the cell positional distribution has its highest values at cell position four. However, apoptotic cells do not appear to express this protein and the strongly p53-positive cells are clearly nonapoptotic, but occur at the same position in the crypt as the apoptotic cells. The fact that we do not see the protein in the apoptotic cells does not, however, preclude the possibility that the p53 protein was involved in the triggering of the apoptotic event. It may have occurred at times that we have not studied or more likely, it occurs at low levels below the threshold of detection using immunohistochemistry. However, there are some strongly p53-positive cells and our interpretation of these is that they are the siblings of the ultimate stem cells, which are going to regenerate the stem cells destroyed via apoptosis and ultimately regenerate the whole crypt (i.e., they are clonogenic stem cells). These cells may be at a cell cycle checkpoint, probably in G2, where they are delayed for a sufficient length of time to ensure that any DNA damage is efficiently repaired. The gene $p21^{WAF1/cip-1}$ is transcriptionally regulated by p53 and is also elevated following radiation exposure over a wider range of cell positions than p53. $p21^{WAF1/cip-1}$ brings about cell cycle arrest by inhibiting the action of the cyclin proteins, which are essential for transition through the various stages of the cell cycle.

Apoptosis induced by other cytotoxics: are all cells programmed to die?

In the mid-1980s we undertook, with a colleague, Kenichi Ijiri, a series of experiments to study the effects of eighteen different cytotoxic drugs, radiation, and chemicals. This was subsequently supplemented by a study by a colleague, Yu Quin Li, looking at four mutagens/carcinogens. For each agent, different doses and different times after treatment were analysed. These data were analysed to obtain information about the position in the crypts at which the cell death was induced. The results were rather surprising in that the cytotoxic agents fell into three broad categories (see figure 46). There were those already discussed in this chapter that targeted cells at around the fourth/fifth position in the crypt. There was a second group of drugs that targeted cells slightly higher up the crypt centred around cell position 7, and these were presumed to act somewhat selectively on cells early in the dividing transit lineage. Then there was a final group of drugs that targeted cells at around the tenth cell position acting predominantly on cells in the mid to late transit populations. These observations were important because they showed that all the cells in the crypt hierarchy could, after appropriate damage stimulation, initiate the apoptosis sequence. We never observed with any of the agents that were studied, at any of the doses, changes that were indicative of necrosis. All of these agents seemed to cause damage that resulted in the triggering of apoptosis. This is consistent with the concept proposed by Martin Raff that all cells possess the programme for cell death but they are prevented from dying by survival signals, which can be clearly overridden, in this case by the appropriate cytotoxic agent. It is difficult, but not impossible, to induce apoptosis in the differentiated compartment of the intestine. There are suggestions that certain bacterial toxins and treatments such as occluding the blood supply precipitate apoptosis on the villus. However, this has not been very extensively studied. As already mentioned, there are also indications that as the cells approach the tip of the villus some of them may initiate and undergo part of the apoptosis sequence prior to extrusion. Both *bax* and *bad* (cell death genes) are expressed

AN *IN VIVO* SYSTEM TO STUDY APOPTOSIS

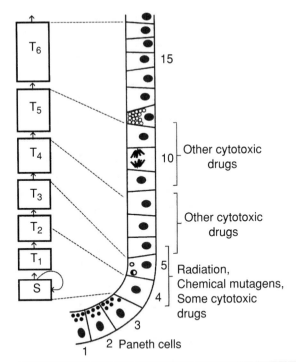

46. Schematic diagram showing the median value for cell positional apoptotic yields in the small intestine, obtained using a wide variety of radiation sources, cytotoxic drugs, and chemical carcinogens. Reproduced with permission from Potten, 1996.

at the villus tip. This is again consistent with the concept that the fallback status of cells is apoptosis. However, many cells are extruded as healthy-looking cells.

Radiation, as already mentioned, is a useful cytotoxic, because the dose can be very precisely controlled and all cells within the focus of the radiation field receive an equal dose. It is the only cytotoxic agent where we know that damage is induced in all cells when we switch the X-ray set on and that damage induction stops when we switch it off. Using the dose–response data shown in figure 45, one can select doses of radiation that will kill on average one cell at the stem cell position in each crypt, or doses that will kill two or three, and so on. A variety of experiments have been performed to look at how the crypt responds to the induction of low levels of cell death. These experiments all showed the same thing, namely that the crypt can detect the death of a single cell in the stem cell region and that it responds by changing the cell proliferation characteristics of the remaining cells at the stem cell position. These changes are almost instantaneous and are presumed to result in the replacement of the cell that has been lost. The conclusions that can be drawn from these radiation experiments and from the studies on spontaneous apoptosis are that the stem cells are continuously monitoring their numbers and if their numbers increase by one, apoptosis is triggered in one of their number. If the numbers decrease by one, proliferation is triggered in one or more of the remaining cells to replace the cell that is lost. This regulatory mechanism ensures that the crypt remains at a constant and stable size, which is clearly important for the tissue. What is fascinating is how this regulation is achieved. What are the mechanisms involved? This is a complex cell-biological problem that has to take into account, amongst other things, the complex geometry of the crypt, which is a flask-shaped structure with the bottom few positions predominantly filled by Paneth cells. As discussed in this chapter, Paneth cells are distributed irregularly at the base of the crypts and consequently, so is the annulus (ring) of cells containing the ultimate stem cells (figure 39). Somehow, despite their nongeometric positioning, stem cells numbers can be determined. This is a fascinating and intriguing problem – how is it achieved? What are the signals and what are the implications? This is one of the challenges for those working in gastrointestinal cell biology over the early twenty-first century.

Apoptosis in the large bowel and the role of Bcl-2

Now, what can we say about the radiation-induced apoptosis in the large intestine? The data here tend to be much more inconsistent and this we think is related to the rather variable expression of the Bcl-2 protein. In some of the earlier experiments, radiation-induced apoptosis occurred at a lower overall frequency for a particular dose and showed no particular cell positional association. At this time, Bcl-2 expression was evident at the base of the crypts in our mice. When *bcl-2* knockout animals were studied, the levels of spontaneous apoptosis

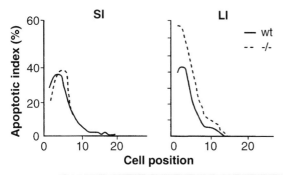

47. Cell positional apoptotic counts in the small and large intestine of normal, wild-type (wt), and *bcl-2* knockout (−/−) mice at three hours following a dose of 8 Gy of radiation. In the small intestine the gene deletion has no effect, whereas in the large bowel the incidence of apoptosis is increased in the knockout animals.

in the stem cell region increased dramatically. When the levels of radiation-induced apoptosis were studied, they too were dramatically increased in the large intestine of *bcl-2* knockout mice (figure 47), whereas no changes were observed in the small intestine. However, subsequent experiments resulted in sporadic confirmation of these observations. As indicated in the preface, nothing in science ever seems to be completely certain. Our conclusions at the moment are that the survival or death of cells in the stem cell region of the colon varies, as a consequence of the differing expression of the various members of the Bcl-2 family, which change for reasons that we do not understand at the moment. When the *bcl-2* gene is deleted, some animals die *in utero* as a consequence, some die shortly after birth, some grow to be adults of a diminished stature (runts), whereas others, which have been genotyped to confirm that they lack the *bcl-2* gene, look perfectly healthy, other than having some grey fur. These mice can breed. It is clear that different members of the *bcl-2* family of genes compensate for the absence of *bcl-2* to differing degrees in these animals. *bcl-2* gene expression may vary in wild-type animals as well. We have tended to breed the *bcl-2* knockout animals from the more healthy individuals and, as a consequence, I think we have selected for animals with a particular constitution of *bcl-2* family genes, which result in variable apoptotic distributions. It is also possible that genes such as *bcl-2* are influenced by other factors in the gut, such as the intestinal bacterial flora, which may vary from animal to animal and from time to time.

Bcl-2 expression in human colon appears to be much more consistent and perhaps is not subject to the same competitive interaction between genes. Bcl-2 tends to be even more strongly expressed in normal looking crypts adjacent to cancers in humans. It is also expressed in the early developing cancers. As the cancers become more aggressive, Bcl-2 tends to be more weakly expressed, whereas p53 in a mutant form becomes more strongly expressed. The conclusions that can be drawn in the colon are as follows. The strong protective mechanism afforded by apoptosis seen in the small intestine does not appear to be as consistently operating in the large bowel, probably because of the action of Bcl-2. The consequences are that cells that have sustained damage to their DNA may successfully repair the damage, or may not detect the damage, or they may misrepair the damage. Consequently, there is a greater risk that damage will persist or be induced by repair, which will result in a higher cancer risk in the colon. Cancers that develop from the Bcl-2-expressing stem cells of the colon may initially continue to express Bcl-2. As a consequence, such cells would be expected to be difficult to kill (difficult to trigger apoptosis) and this may account for the high chemoresistance of tumours in the large bowel. However, it is clear that as the tumour progresses Bcl-2 expression affords less advantage to the tumour, relative to the expression of mutated p53 protein, which will disable the initiation of arrest and repair or apoptosis programmes, in response to DNA damage (i.e., chemotherapy). Interestingly, we have shown that a homologue of Bcl-2, named Bcl-w or Bcl-2L2, is expressed in advanced tumours that do not express Bcl-2 and do possess mutated p53.

It is unclear why the small and large bowel should have adopted different strategies for dealing with DNA damage (the 'better dead than damaged' strategy in the small bowel but 'better damaged than dead' in the large bowel, or 'better dead than red' or 'better red than dead' to coin expressions from the days of the Cold War!). One possible explanation is that in the small bowel, DNA-damaging agents are rare and the 'better dead' strategy provides an absolute protective mechanism. Here, occasional cell suicide can be fairly safely compensated for by occasional, symmetric stem cell divisions. In the large bowel, in contrast, toxic chemicals (mutagens) might be encountered more frequently and the 'better dead' strategy would result in constant minor regenerative proliferation. This carries its own risks of DNA replication-induced errors and here the 'better alive' strategy combined with efficient repair would be the desired approach.

The *adenomatous polyposis coli* (*APC*) gene

It has been known for a long time that several genes need to be mutated in order for a normal cell to become cancerous (see figure 48). One of the most important is the *adenomatous polyposis coli* (*APC*) gene. The APC protein is responsible for directing the degradation of free β-catenin within the cell. As mentioned previously, β-catenin may combine with E-cadherin to promote cell-to-cell adhesion; alternatively it may act to regulate gene expression required to drive cell proliferation. Obviously, this latter activity needs to be carefully regulated. If APC function is blocked by mutations that either inactivate the protein or result in its nonexpression, cellular β-catenin levels will be unchecked and dysregulated proliferation may result. This may ultimately lead to neoplasia and eventually malignant cancer. Many tumours of the large bowel show elevated levels of β-catenin in cell nuclei and mutation in the *APC* gene.

A second area of current interest is the cyclooxygenase enzymes (COX1 and COX2), which are involved in the conversion of arachadonic acid into prostaglandins and which have a direct effect on proliferation (not to be confused with the cytochrome oxidase enzymes in the mitochondria, which are also referred to as COX and are also implicated in apoptosis). COX1 is the normal constitutive enzyme, whereas COX2 tends to be induced in abnormal states, for example, in advanced tumours of epithelial tissues such as the large bowel (adenocarcinomas). The importance of elevated COX levels may be that increased levels of prostaglandins induce the expression

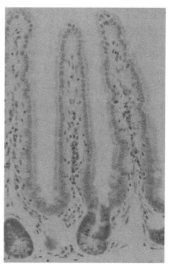

```
         NORMAL
APC ──────▶
    EGF   MICROADENOMA
    ──────▶
    K-Ras   EARLY ADENOMA
    ──────▶
    TGF-β   MID ADENOMA
    ──────▶
    p53    LATE ADENOMA
    ──────▶
            ADENOCARCINOMA
```

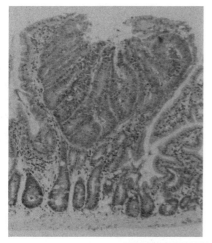

48. Diagram showing the sequence of gene mutations required to change a normal intestinal stem cell into a cancer cell. *APC* plays a critical role in this sequence and is sometimes referred to as the 'gatekeeper' gene. The inserts show normal intestine (top right) and an adenoma in the small intestine of mice with a mutated *APC* gene.

of Bcl-2 and, therefore, increase the resistance of tumour cells to apoptosis. Both these enzymes are inhibited by nonsteroidal anti-inflammatory drugs (NSAIDS), such as aspirin and sulindac, both of which have been implicated in reducing colon cancer risk, as well as facilitating in the cure of existing tumours. In addition to COX inhibition, these drugs may also have a direct proapoptotic effect on tumour cells.

Small and large intestine cancer incidence figures

One of the surprising and poorly understood features of the gastrointestinal tract is its unusual cancer incidence distribution. There has been no completely adequate explanation for the fact that in the large intestine, particularly in certain regions of the large intestine, the cancer incidence figures in the western world are very high, in fact cancer of the colon is the second most common form of cancer. Cancers of the colon are also difficult to cure, the tumours are very resistant, and patients with colon cancer tend to have a rather poor prognosis and, hence, high mortality rate. Cancer is also fairly common in the stomach and the oesophagus and the oral cavity, but rare in the small intestine.

The small intestine in both mouse and humans is between 3 and 4 times greater in terms of total mass (length) than in the large intestine. The stem cells in the small intestine are between 1.7 and 2.5 times greater in number in the small intestine than colon and they divide about 1.5 times more in terms of total number of divisions in the small intestine than the large intestine, with a cycle time that is about 1.5 times faster in the small intestine than the large intestine. Bearing in mind these numbers, all of which would suggest a greater risk of accumulating the necessary genetic errors for the development of cancer in the small intestine, it is extremely surprising that the cancer incidence figures here are about seventyfold lower than they are in the large intestine. In fact, in some regions of the small intestine cancer is almost unknown. Although many attempts have been made to answer the question, 'What are the factors that predispose colonic stem cells to carcinogenic transformation?' few have attempted to answer the reciprocal question, 'What is it that protects the small intestine so effectively?' As just discussed in the preceding section, it is probable that apoptosis plays a key role in protecting, at least in part, the small intestine against the establishment of clones of cells carrying DNA carcinogenic lesions.

Genome protection mechanisms

What the comparison of cancer incidence figures in the gut suggests is that the stem cells in the small intestine have evolved extremely powerful and effective mechanisms to protect against the accumulation of genetic errors. John Cairns, in a controversial article in *Nature* in 1975, suggested that the most genetically hazardous event in the life cycle of a cell is DNA replication. This is the stage in the cell cycle when there is the greatest risk of incorporating an incorrect base and, hence, inducing a mutation. He further suggested that stem cells would have evolved a variety of protective mechanisms and that a very effective and simple mechanism to completely eliminate the risk of DNA replication-induced errors would be for the cells to have evolved a mechanism for sorting the old from the new DNA strands at mitosis. The stem cells would retain the old strands that were used as templates in the last round of DNA synthesis and discard the newly synthesised strands to the daughter cells that were going to enter the dividing transit population, differentiate and be shed from the tissue a relatively short time later.

Although we generated some data that provided circumstantial support for this hypothesis in 1978, it has been difficult to provide convincing proof or disproof; however, we have recently generated some data for the small intestinal stem cells that clearly indicate that a selective strand segregation process is operating. These studies involved careful labelling of the template strands at times when stem cells are making new stem cells and, hence, making new templates and this results in the permanent labelling of the stem cells. These studies were combined with subsequent labelling with a different DNA marker at a subsequent round of DNA synthesis. The fate of the two markers was then studied over a range of times that spanned many rounds of divisions. It was found that the two markers segregated with the passage of time in a totally different fashion. The initial marker incorporated into the template strands (tritiated thymidine) remained permanently in the cells presumed to be the stem cells, whereas the marker incorporated into the newly synthesised strands (bromodeoxyuridine) disappeared at a time equivalent to a second division of the stem cells (see figure 49).

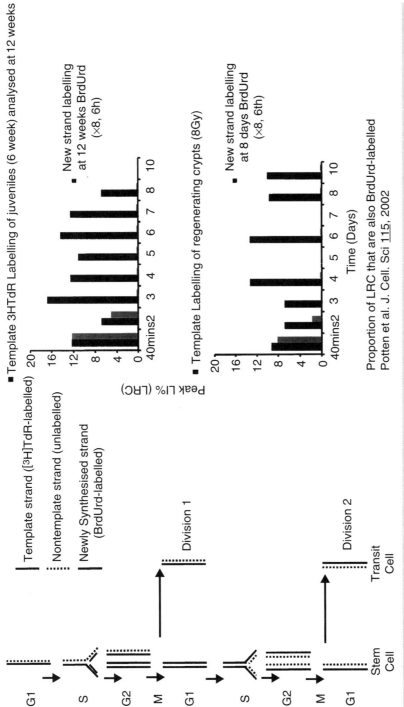

49. Diagram showing the selective DNA strand segregation hypothesis, where the template strands are retained in the stem cell and the newly synthesised strands, with possible replication-induced errors, are passed to the daughter cell of an asymmetric stem cell division, destined to be a transit cell and lost to the system after a few days. The bar diagram shows the results of an experiment where the template strands were labelled with tritiated thymidine and the newly synthesised strands with bromodeoxyuridine. The data presented are for cells at the stem cell position in the crypts and show that template labelling persists, whereas the new strand labelling disappears at around the second day (at the second stem cell division).

A further assumption within the Cairns's hypothesis was that the stem cells that were selectively segregating their DNA strands would operate a prohibition of processes like sister chromatid exchange, because this would then result in a mixing of the template and newly synthesised strands. This would suggest that these cells would also lack excision repair capabilities, because several of the enzymes in sister chromatid exchange and excision repair are common. An inability to undergo repair would imply that the cells were exquisitely sensitive and we now believe that this is the explanation for the extreme sensitivity exhibited by four to six cells at the stem cell position in the crypt. These cells do not apparently possess a repair capacity but, once damage has occurred in their DNA, activate an altruistic cell suicide or apoptosis. This provides a very effective second level of protection against random genotoxic damage in the template strands or against double-strand errors that also involve the template strands. The random genotoxic events could arise from background radiation or the presence of genotoxic chemicals generated by the bacterial flora in the intestine, although this is likely to be rather limited in the small intestinal regions.

The induction of the altruistic suicide mechanism in stem cells by random genotoxic damage is dealt with by the fact that not all the stem cells will be killed, but if many are killed as in an experimental situation (but a situation that rarely occurs in nature), then the small intestinal crypts have a hierarchy of stem cells with second- and third-tier cells that possess good repair capacity, a great radio resistance (genotoxic resistance), and an ability to repopulate the stem cell compartment and the entire cell lineages should the ultimate stem cells all be deleted. Thus, the system is extremely well protected (see figure 50).

The DNA segregation studies have not been performed in the large bowel so we do not know whether this mechanism is compromised in this region of the gut that develops many cancers. However, we do know that the altruistic cell suicide process is compromised in a fairly major way by the expression of the protein derived from the *bcl2* gene, which is an antiapoptotic cell survival gene. This protein is not expressed in the stem cells of the small intestine but it is

AN *IN VIVO* SYSTEM TO STUDY APOPTOSIS

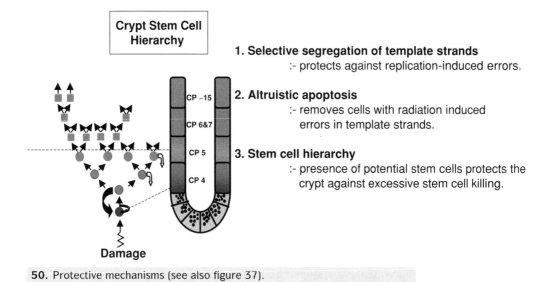

50. Protective mechanisms (see also figure 37).

expressed in the stem cell regions of the large bowel in both mouse and humans. Compromising one of the protective mechanisms may be sufficient to account for the increased cancer risk. However, experiments are ongoing to determine whether the strand segregation process is similarly compromised.

Recent studies by Sherry at MIT in Boston, using cell culture systems where the *p53* gene has been genetically manipulated such that it is under the control of a zinc-dependent promoter gene and so can be switched on and switched off in the cells by the simple addition or removal of zinc, have provided some interesting further information on the DNA segregation story. When p53 is absent the cells divide in a symmetric fashion, both daughters of a division being indistinguishable. However, the switching on of *p53* by the presence of zinc results in the cells adopting an asymmetric division mode where one daughter remains a proliferative cell and the other differentiates. Thus, p53 seems to be involved in controlling some aspects of symmetric and asymmetric divisions. The asymmetric division mode is reminiscent of the processes expected of stem cells in tissues under steady-state conditions. Using ingenious labelling protocols, Sherry was also able to show that the asymmetric division mode was further associated with an asymmetric segregation process of the DNA strands; this provides a much more complex role for p53 than simply controlling the decision as to whether a cell will enter cell cycle arrest or initiate apoptosis. In the small intestinal stem cells, an hypothetical model of p53 action can be proposed. In the stem cells p53 controls the altruistic apoptotic response in the stem cells. These cells do not appear to have the option to undergo cell cycle arrest and repair. This is reserved for cells within the dividing transit population, so the distinction between these two pathways for p53 seems to be associated with stem and transit cells respectively. If the dose of radiation is increased, experimental data and mathematical modelling suggest that many cells undergo a permanent cell cycle arrest that may also be dependent on the p53–p21 pathway, although this has not been proven. What the Sherry article seems to show is that p53 is also controlling asymmetric divisions, which are divisions associated with the stem cells and the selective strand segregation process. So, these processes can be put together in the form of a diagram or flow chart of the sort shown in figure 51.

The DNA segregation story has major implications in cancer cell biology and cancer genetics as well as in studies into ageing. One mechanism that has been proposed for ageing is that the DNA strands have specialised end regions called telomeres and that telomeres have a problem being adequately replicated at each round of DNA synthesis, a process that can be overcome by using a special enzyme called telomerase, which is switched on in cells such as cancer cells. However, if selective DNA strand segregation occurs the template strands would not be subject to these end replication problems and the telomeres would not shorten in stem cells.

In invertebrate as well as mammalian cell systems there are a number of genes and their proteins that have been associated with asymmetric division processes. These include the membrane-associated gene products, notch and delta, and the intracellular cytoplasmic proteins produced by the numb gene. Some of these genes are illustrated in a simple way in figure 52.

Although there are no ways of marking intestinal stem cells the selective DNA strand mechanism does offer a means of permanently labelling these cells. If the template strands can be labelled when stem

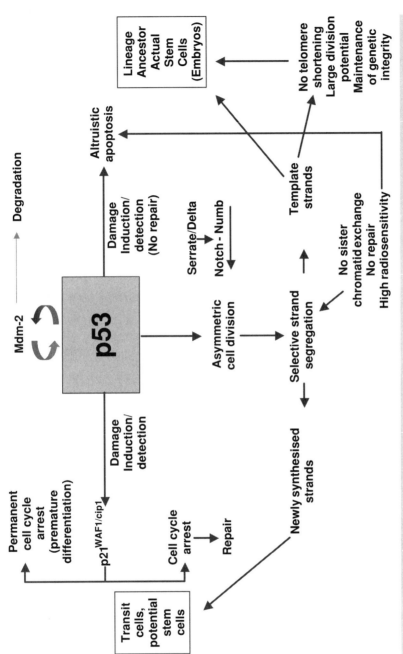

51. Diagram showing the hypothetical role for p53 in the stem and transit populations of the small intestinal crypt. In the stem cells, p53 appears to trigger apoptosis and does not permit cell cycle arrest and repair. The converse is true for the transit cells, which seem to repair but do not undergo apoptosis. Evidence is accumulating that p53 may also regulate asymmetric cell division in stem cells and be involved in DNA strand segregation; two processes specifically associated with stem cells. Modified from Potten et al., 2003.

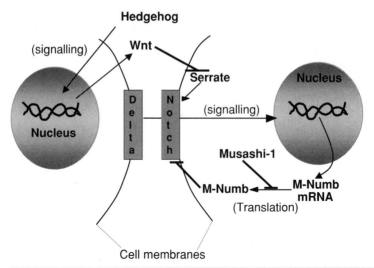

52. A probable scheme of interacting genes controlling asymmetric divisions in stem or progenitor cells, particularly in some invertebrate systems during development.

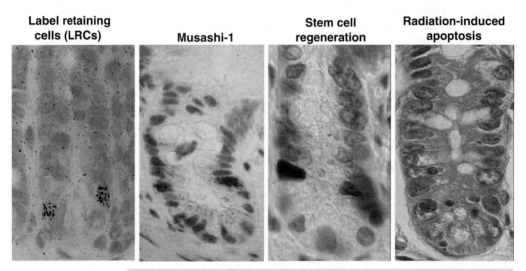

53. Four approaches to studying small intestinal stem cells. Labelling the template strands according to the Cairns's hypothesis generates label-retaining stem cells; antibody staining to *musashi-1* identifies early lineage cells (including stem cells) and the radiation-induced apoptosis is a response specifically related to the stem cells. The early stages of postinjury regeneration are believed to involve stem cells. Twenty-four hours after a high dose of a chemotherapeutic agent, a few cells at the stem cell position in the crypt enter DNA synthesis and can be labelled with bromodeoxyuridine.

cells are being made, those cells will then have permanent labelled DNA. The difficulty here is that this type of labelling involves the use of radioactive DNA precursors and the process of autoradiography, but it does offer a highly specific method for labelling stem cells. It turns out that one of the genes associated with the delta-notch-numb pathways is expressed in early lineage cells in the intestine. This gene, called *Musashi-1 (Msi-1)*, interferes with the translation of numb protein. Named after a famous Japanese samuri who wielded two swords, *Msi-1* is a gene known to be involved in early asymmetric cell divisions in neural stem cells in both invertebrates and mammals. Using antibodies, this Msi-1 protein can be seen to be expressed in a few cells at the stem cell position and shows variations in expression levels in situations where the stem cell numbers would be predicted to vary such as during development, postradiation regeneration, and cancer development. In a rather indirect way another way of identifying the presence of these stem cells is by virtue of their ability to undergo altruistic apoptosis. The similarity in these three processes is illustrated in figure 53.

Another possible way to see the stem cells is to observe them performing one of their functions: tissue regeneration. It has been known from many experiments using various cytotoxic insults that the initial phases of regeneration are broadly associated with the lower regions of the crypt. Recently we have observed that following a high dose of a drug called 5-fluorouracil, all cell proliferation ceases but at twenty-four hours a few cells at cell position four enter the S phase, presumably to initiate the regenerative response.

Summary

In the small intestine of the mouse the yield of apoptosis in various circumstances shows some relationship to the levels of mitotic activity, suggesting that these two processes are in some way linked or interrelated. Generally when proliferation is elevated so is apoptosis.

One of the things that one has to remember with the interpretation of such studies is that the genes that regulate apoptosis are numerous and varied. The *bcl-2* family contains many genes and the number

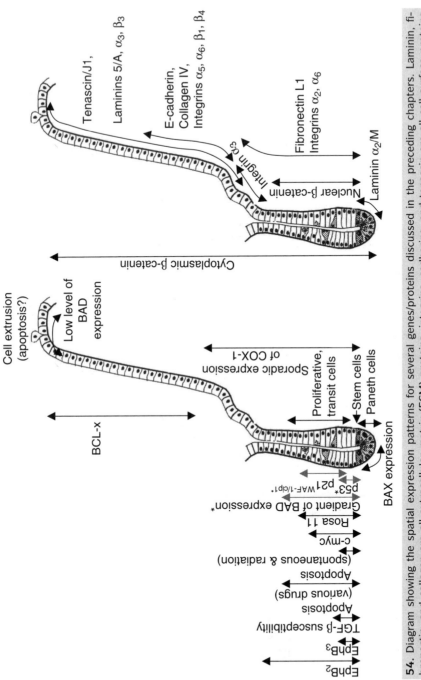

54. Diagram showing the spatial expression patterns for several genes/proteins discussed in the preceding chapters. Laminin, fibronectin, and collagen are all extracellular matrix (ECM) proteins; integrins, cadherins, and tenascin are all cell surface proteins involved in cell-cell or cell-ECM adhesion. ROSA-11 is a marker for early lineage cells. Modified from Potten, 1998.

AN *IN VIVO* SYSTEM TO STUDY APOPTOSIS

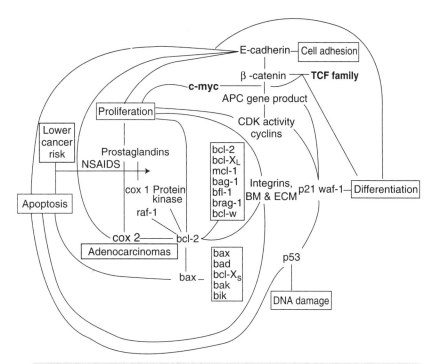

55. Highly schematic diagram showing probable interactions between some of the genes outlined in the previous two chapters. The lines indicate only some level of interaction, not up or down regulatory processes, but are shown to illustrate the complexity of the processes being studied. Modified from Potten, 1998.

of members that are discovered increase regularly. There are also a whole series of other gene families, including those coding for the caspases and CARD proteins, which play a major role in regulating apoptosis. In order to understand the regulatory processes and the role of these genes, we probably need to know something about the expression levels of most or all of them. At present, in the intestine, the information is restricted to only a small number. A summary of the distribution of some of the other *bcl-2* family genes, together with information on the spatial distribution of apoptosis produced under various circumstances and the expression of basement membrane and extracellular matrix proteins, is shown in figure 54. This complex figure is not intended to confuse, but to give some indication of the complexity of the system and the multitude of genes, factors, and proteins that may play important roles in determining the environment for stem cells, which itself may determine how effectively they function, whether they proliferate, differentiate, or die. This is probably only a small proportion of the information that is required before we fully understand these regulatory processes. Figure 55 attempts to give a further level of complexity to the system by indicating some of the known, or suspected, interactions among genes, enzymes, basement membrane proteins, and extracellular matrix proteins and their possible involvement in proliferation, differentiation, and apoptosis. Some of these pathways and interactions are the subject of intense research activity at present and could lead not only to an understanding of the regulatory processes that I have been discussing, but also to the changes that lead to cancer and the possible evolution of new strategies for cancer treatment and cancer prevention.

Further reading

Battle, E., Henderson, J. T., Beghtel, H., van den Born, M. M. W., Sancho, E., Huls, G., Meeldijk, J., Roberston, J., van de Wetering, M., Pawson, T., and Clevers, H. β-catenin and TCF mediate cell positioning in the intestinal epithelium by controlling the expression of EphB/Ephrin B. *Cell* **111**: 251–263, 2002.

Bjerknes, M., and Cheng, H. Modulation of specific intestinal epithelial progenitors by enteric neurons. *Proc. Natl. Acad. Sci. USA* **98**: 12497–12502, 2001.

Bjerknes, M., and Cheng, H. Clonal analysis of mouse intestinal epithelial progenitors. *Gastroenterology* **116**: 7–14, 1999.

Dünker, N., Schmitt, K., Schuster, N., and Krieglstein, K. The role of transforming growth factor beta-2, beta-3 in mediating apoptosis in the murine intestinal mucosa. *Gastroenterology* **122**: 1364–1375, 2002.

Hagios, C., Lochter, A., and Bissell, M. J. Tissue architecture; the ultimate regulator of epithelial function. *Phil. Trans. R. Soc.* (B) **353**: 857–870, 1998.

Hall, P. A., Coates, P. J., Ansari, B., and Hopwood, D. Regulation of cell number in the mammalian gastrointestinal tract: the importance of apoptosis. *J. Cell Sci.* **107**: 3569–3577, 1994.

Hendry, J. H., Potten, C. S., Chadwick, C., and Bianchi, M. Cell death (apoptosis) in the mouse small intestine after low doses: effects of dose rate, 14.7 Me V neutrons and 600 Me V (maximum energy) neutrons. *Int. J. Rad. Biol.* **42**: 611–620, 1982.

Hermiston, M. L., and Gordon, J. I. Inflammatory bowel disease and adenomas in mice expressing a dominant negative N-cadherin. *Science* **270**: 1203–1207, 1995.

Hermiston, M. L., Wong, M. H., and Gordon, J. I. Forced expression of E-cadherin in the mouse intestinal epithelium slows cell migration and provides evidence for nonautonomous regulation of cell fate in a self-renewing system. *Genes Dev.* **10**: 985–996, 1996.

Ijiri, K., and Potten, C. S. Further studies on the response of intestinal crypt cells of different hierarchial status to cryptoxic drugs. *Br. J Cancer* **55**: 113–123, 1987.

Li, Y. Q., Fan, C., O'Connor, P. J., Winton, D., and Potten, C. S. Target cells for the cytotoxic effects of carcinogens in the murine small bowel. *Carcinogenesis* **13**: 361–368, 1992.

Marshman, E., Booth, C., and Potten, C. S. The intestinal epithelial stem cell (our favourite cell). *Bioessays* **24**: 91–98, 2002.

Merritt, A. J., Potten, C. S., Kemp, C. J., Hickman, J. A., Balmain, A., Lane, D. P., and Hall, P. A. The role of the p53 in spontaneous and radiation induced apoptosis in the gastrointestinal tract of normal and p53 deficient mice. *Cancer Res.* **54**: 614–617, 1994.

Merritt, J. A., Potten, C. S., Watson, A. J. M., Loh, D. Y., Nakayama, L., Nakayama, K., and Hickman, J. A. Differential expression of bcl-2 in intestinal epithelial: correlation with attenuation of apoptosis in colonic crypts and the incidence of colonic neoplasia. *J. Cell. Sci.* **108**: 2261–227, 1995.

Mills, J. C., and Gordon, J. I. The intestinal stem cell niche: there grows the neighborhood. *Proc. Natl. Acad. Sci. USA* **98**: 12334–12336, 2001.

Potten, C. S., Li, Q. Y., O'Connor, P. J., and Winton, D. J. A possible explanation for the differential cancer incidence in the intestine, based on the distribution of the cytotoxic effects of carcinogens in the murine large bowel. *Carcinogenesis* **13**: 2305–2312, 1992.

Potten, C. S., and Hendry, J. H. (eds.). Cytotoxic Insult to Injury. Churchill–Livingstone, Edinburgh, p. 421, 1983.

Potten, C. S. What is an apoptotic index measuring? A commentary. *Br. J. Cancer* **74**: 1743–1748, 1996.

Potten, C. S. The significance of spontaneous and induced apoptosis in the gastrointestinal tract of mice. *Cancer Metastasis Rev.* **11**: 179–195, 1992.

Potten, C. S. Structure, function and poliferative organisation of mammalian gut. In: Radiation and Gut, C. S. Potten and J. H. Hendry (eds.). Elsevier, Amsterdam, pp. 61–84, 1995.

Potten, C. S., Wilson, J. W., and Booth, C. Regulation and significance of apoptosis in the stem cells of the gastrointestinal epithelium. *Stem Cells* **15**: 82–93, 1997.

Potten, C. S., Booth, C., and Hargreaves, D. The small intestine as a model for evaluating adult tissue stem cell drug targets. *Exp. Med.* **21**: 59–68, 2003 (in Japanese). English version: Cell Prolif. **36**: 115–129, 2003.

Potten, C. S. Stem cells in gastrointestinal epithelium: numbers, characteristics and death. *Phil. Trans. R. Soc. Lond, (B)* **353**: 821–830, 1998.

Potten, C. S. Structure, function and proliferative organisation of the mammalian gut. In: Radiation and Gut, C. S. Potten and J. H. Hendry (eds.). Elsevier, Amsterdam, pp. 1–31, 1995.

Potten, C. S. A comprehensive study of the radibiological response of murine (BDFI) small intestine. *Int. J. Radiat. Biol.* **58**: 925–973, 1990.

Potten, C. S., and Grant, H. The relationship between radiation-induced apoptosis and stem cells in the small and large intestine of mice. Br. J. Cancer, **78**: 993–1003, 1998.

Potten, C. S., and Hendry, J. H. Clonal regeneration studies. In: Radiation and Gut, C. S. Potten and J. H. Hendry (eds.). Elsevier, Amsterdam, pp. 45–59, 1995.

Potten, C. S. Extreme sensitivity of some intestinal crypt cells to X and γ irradiation. *Nature* **269**: 518–521, 1977.

Potten, C. S., Booth, C., and Pritchard, D. M. The intestinal stem cell: the mucosal governor. *Int. J. Exp. Pathol.* **78**: 219–243, 1997.

Stappenback, T. S., Wong, M. H., Saam, J. R., Mysorekar, I. U., and Grodon, J. I. Notes from some crypt watchers: regulation of renewal in the mouse intestinal epithelium. *Curr. Opin. Cell Biol.* **10**: 702–709, 1998.

Taipale, J., and Beachy, P. A. The Hedgehog and Wnt signalling pathways in cancer. *Nature* **411**: 349–354, 2001.

Walters, J. R. F. New advances in the molecular biology of the small intestine. *Curr. Opin. Gastroenterol.* **18**: 161–167, 2002.

van de Wetering, M., Sancho, E., Verweij, C., de Lau, W., Oving, O., Hurlstone, A., van der Horn, K., Battle, E., Coudreuse, D., Haramis, A.-P. *et al*. The β-catenin/TCF complex imposes a crypt progenitor phenotype on colorectal cancer cells. *Cell* **111**: 241–250, 2002.

Wong, M. H., Huelsken, J., Birchmeier, W., and Gordon, J. I. Selection of multipotent stem cells during the morphogenesis of small intestinal crypts of Leiberkühn is perturbed by stimulation of Lef-1/β-catenin signalling. *J. Biol Chem.* **277**: 15843–15850, 2002.

Wong, M. H., Rubinfield, B., and Gordon, J. I. Effects of forced expression of an NH_2-terminal truncated β-catenin on mouse intestinal epithelial homeostasis. *J. Cell Biol.* **141**: 765–777, 1998.

Wright, N. A., and Alison, M. The Biology of Epithelial Cell Populations, Vol. 2. Clarendon Press, Oxford, 539–1247, 1984.

8

Cell death (apoptosis) in diverse systems

So far in this book we have examined the role that apoptosis plays during both mammalian development and adult life. We have also looked at apoptosis in a simple organism, the nematode *C. elegans*, and discussed how information from studies on this organism has helped our understanding of the regulation and execution of apoptosis in mammals.

Apoptosis has now been reported in almost all other animal systems studied ranging from simple organisms, including coelentrates (which posses just two cell layers) such as hydra, annelids such as leaches, insects such as moths, grasshoppers, and fruit flies, amphibia, and fish, in particular the zebrafish, which was already a favourite experimental model for developmental biologists. It has also been described in plants. Good examples of the crucial importance of apoptosis in these other organisms are during metamorphosis of insects from larval to adult forms, and the metamorphosis of amphibia from water-living tadpoles to air-breathing adults. One example is discussed below.

A classic and often-quoted example where the organised removal of cells is required during the course of development is the absorption of the tail of a tadpole. When a tadpole changes into a frog the tail does not fall off but is gradually 'absorbed' by the developing frog. This apparent 'absorption' is achieved via apoptosis of the tail cells. Of the various tissues that make up the tail, the cells in the epidermis can be seen to undergo the classical morphological changes associated with apoptosis. The cells die, showing cytoplasmic condensation, chromatin condensation, fragmentation, and engulfment of the

CELL DEATH (APOPTOSIS) IN DIVERSE SYSTEMS

fragments. The removal of the dying cells is predominantly achieved by a 'professional' debris remover, the macrophage. These cells can be seen to contain large numbers of apoptotic fragments and large numbers of whorled structures (myelin whorls) that are believed to represent the remains of apoptotic fragments. A few of the fragments are engulfed by healthy epithelial cells, but the bulk of the removal is achieved by these macrophages.

The muscles in the tail are made up of striated muscle cells. These are large cellular structures and one might not expect them to undergo the same rounding up and fragmentation that is seen for small individual cells. However, the muscle cell nuclei show nuclear changes consistent with apoptosis and the muscle fibres fragment into specialised apoptotic bodies (sarcolytes) and again these are removed by macrophages. The interesting questions raised by this system concern how this process is regulated: How do the cells in tadpole tail know when to initiate the process of apoptosis? And how do they know where and when to stop? It is clear in this system that systemic hormones play an important role, particularly those derived from the thyroid gland, which is dependent on iodine.

Like the tadpole, apoptosis is crucial in mammalian development, controlling body patterning and organ, limb, and digit formation. Failure to regulate apoptosis precisely during development can manefest itself as an observable phenotype. Cleft palate appears to be due to excessive apoptosis of the ectomesenchymal cells of the palate during craniofacial development. In contrast, reduced apoptosis of mesodermal cells between the limb digits can result in webbed fingers or toes.

Within adult mammals, there is a range of diverse tissue systems, all with their own specific controls for apoptosis. Good examples of this are the hormone-dependent tissues of the body such as the breast, uterus, and prostate. Androgen removal or decline can induce dramatic regression of the prostate gland, which is achieved via apoptosis. Some tumours of these tissues continue to be steroid hormone-dependent and as a consequence are susceptible to antisteroidal therapies.

The normal breast undergoes cyclical patterns of proliferation and apoptosis under control of the monthly cycle of varying oestrogen levels. Oestrogen stimulates the proliferation of the breast epithelial cells and results in expansion of the ductal network within the breast that produces and transports milk. These changes are in preparation for pregnancy and lactation. Many of the new cells produced are dependent on the maintenance of oestrogen levels for their survival and when these fall, towards the end of the menstrual cycle if pregnancy does not occur, the cells die by apoptosis. In the absence of pregnancy such changes are small; however, following successful implantation of a fertilised egg there is a much greater increase in breast epithelial cell proliferation supported by an increased and maintained level of oestrogen. Following weaning, levels of oestrogen fall and there is increased apoptosis and tissue remodelling. The uterus shows similar cyclical patterns of proliferation and apoptosis, in response on changing oestrogen levels.

The cells responsible for sperm production in the testis (the spermatogonia, spermatocytes and spermatids) are all susceptible to cell death. Spermatogenesis is organised on a cell lineage basis, as illustrated in figure 31. In fact, it represents the longest cell lineage in mammalian systems, which means that the greatest number of functional end cells (spermatozoa) are produced, in theory, from each stem cell division. However, there is considerable cell death that occurs naturally within the lineage, thus reducing the total cell output from each stem cell division. Spermatogonia types A_{2-4} are particularly prone to substantial levels of cell death that are morphologically of the apoptotic type. Interestingly, spermatocytes and spermatids also normally undergo cell death, but this has a necrotic morphology, although it does not induce an inflammatory reaction. This is a rare example of necrosis occurring naturally in a highly proliferative tissue.

This naturally occurring spermatogonial apoptosis and late-lineage necrosis is probably associated with the genetic screening that would be expected in such a tissue to remove genetically defective cells. The cells of testis are also exteremely sensitive to radiation, cytotoxic drugs, hyperthermia, and ischaemia. All of these induce apoptosis in the spermatogonia, but may induce a necrotic type of death in spermatocytes and spermatids. The apoptosis induced by such treatments commonly occurs in clusters of cells. The spermatogonial epithelium is an interesting tissue in that the spermatogonia form syncytia. A syncytium is a continuous cytoplasm with many cell nuclei. In the testis the cytoplasm has narrow bridges, which form a continuous link. It is thought the clustering of apoptosis is due in part to the fact that cell death induced in one nucleus triggers the death of all the nuclei in the syncytium.

The testis is also extremely sensitive to mild increases in temperature, which is why in many mammals the testes are held exterior to the body, in the scrotum, where the temperature can be 2°–7°C below core body temperature. Raising the temperature to 42°–43°C for a short time can induce a wave of apoptosis in the spermatogonia and necrosis in the later stages of the cell lineage.

Within the immune system, apoptosis plays an important role in limiting inflammation and autoimmunity and promoting tolerance to commonly encountered antigens (i.e., those derived from proteins in the diet). During development and maturation of T cells in the thymus (hence, their name), the cells are constantly exposed to self-antigens. T cells that recognise self-antigens (autoreactive T cells) should undergo apoptosis in response to repeated stimulation with the self-antigen they recognise; a pathway dependent on the proapoptotic Bcl-2 protein Bim. Some may have an alternative fate, alive but nonresponsive to self-antigen challenge (a state called anergy), or alternatively they become memory regulatory T cells and actively suppress inflammatory responses to self-antigen. A similar mechanism exists in the peripheral lymphoid organs, the largest of which is the intestine, which contains many specialised aggregates of lymphoid cells. The gastrointestinal lymphoid tissue (GALT) is probably involved in regulating some aspects of epithelial cell production and function through the secretion of cytokines such as interleukins and interferons. In these aggregates (the tonsils of the mouth and Peyer's patches of the small intestine), T and B lymphocytes are constantly presented with antigens that are sampled from the environment, in this case the gastrointestinal tract. Constant exposure to common antigens should induce apoptosis, anergy, or a regulatory fate in T cells that recognise them – this is termed *peripheral tolerance* (as opposed to the central tolerance to self-antigens that occurs in the thymus). This is a clear example of how apoptosis regulates the function of an important body system.

Further reading

Allan, D. J., Harmon, B. V., and Kerr, J. F. R. Cell death in spermatogenesis. In: Perspectives in Mammalian Cell Death, C. S. Potten (ed.). Oxford University Press, Oxford, pp. 229–258, 1987.

Bouillet, P., Purton, J. F., Godfrey, D. I., Zhang, L.-C., Coultas, L., Puthalakath, H., Pellegrini, M., Cory, S., Adams, J. M., and Strasser, A. BH3-only Bcl-2 family member Bim is required for apoptosis of autoreactive thymocytes. *Nature* **415**: 922–926, 2002.

Chen, Y., and Zhao, X. Shaping limbs by apoptosis. *J. Exp. Zool.* **282**: 691–702, 1998.

Ellis, R. E., Yuan, J., and Horvitz, H. R. Mechanisms and functions of cell death. *Ann. Rev. Cell Biol.* **7**: 663–98, 1991.

Kerr, J. F. R., Harmon, B., and Searle, J. An electron-microscopic study of cell deletion in the anuran tadpole tail during spontaneous metamorphosis with special reference to apoptosis of striated muscle fibres. *J. Cell Sci.* **14**: 571–85, 1974.

Metcalfe, A. D., Gilmore, A., Klinowska, T., Oliver, J., Valentijn, A. J., Brown, R., Ross, A., MacGregor, G., Hickman, J. A., and Streuli, C. H. Developmental regulation of Bcl-2 family protein expression in the involuting mammary gland. *J. Cell Sci.* **112**: 1771–1783, 1999.

Neurath, M. F., Finotto, S., Fuss, I., Boirivant, M., Galle, P. R., and Strober, W. Regulation of T-cell apoptosis in inflammatory bowel disease: to die or not to die, that is the mucosal question. *Trends Immunol.* **22**: 21–26, 2001.

Potten, C. S., Watson, R. A., Williams, G. M., Tickle, S., and Howell, A. Cell proliferation in normal human breast. I. The effect of age and menstrual cycle upon proliferative activity. *Br. J. Cancer* **58**: 163–170, 1988.

Strange, R., Li, F., Saurer, S., Burkhardt, A., and Fris, R. R. Apoptotic cell death and tissue remodelling during mouse mammary gland involution. *Development* **115**: 49–58, 1992.

9

Measuring the levels of cell death (apoptosis)

In a replacing tissue such as the intestinal epithelium, there is a direct relationship between the rate of proliferation and the levels of apoptosis. If proliferation is raised as a result of some treatment, or during the course of the natural physiology of the tissue, the number of apoptotic bodies is usually also raised. Conversely, if proliferation is suppressed for one reason or another, the number of apoptotic bodies generally decreases. This suggests an intimate relationship between cell division and cell death, both processes contributing equally to the homeostatic mechanisms that regulate the total cell production and cellularity of the tissue. There is clearly, therefore, a considerable interest in being able to measure the levels of apoptosis (i.e., the amount of cell death that occurs in a tissue or tumour). However, there are a number of difficulties that one encounters when one attempts to measure apoptosis.

The number of cells in a tissue that are proliferating is generally determined by counting an index of proliferation. Commonly, this is achieved by counting the number of mitotic figures, which are relatively easy to recognise, or the number of cells that are replicating their DNA (i.e., in the DNA synthesis phase or S phase of the cell cycle) by incorporating labelled DNA precursors into the DNA that is being synthesised (these methods have been discussed in Chapters 2 and 4). Using these approaches, the number of mitotic figures or labelled cells can be counted relative to the total number of cells in the tissue. The number of mitotic cells compared with the number of labelled DNA-synthesising cells will be in direct proportion to the relative durations of the mitotic phase of the cell cycle to the DNA synthesis phase of the cycle. Generally speaking, the duration of these phases is fairly well established, with mitosis lasting between thirty minutes to one hour and DNA synthesis taking six to twelve hours (commonly around 7). The problem comes with determining the denominator in such an index. Ideally, the denominator should be the number of cells that are cycling (i.e., actively proliferating). Unfortunately, not all of these cells are recognised and so an approximation is used. For example, in the epidermis proliferation is restricted to cells in the basal layer and in the intestine the crypts are the proliferative units. However, in the crypt, which contains about 250 cells in total, only 150 are probably actively proliferating. In the basal layer of the epidermis this proliferating fraction, or growth fraction, is not well defined but is probably also about 0.6.

With apoptosis, the determination of the denominator becomes more complicated because one has virtually no idea about the size of the apoptosis-susceptible population in the tissue. If we take the intestinal epithelium as the example and consider sections cut through the epithelium along the crypt–villus axis, we may see one apoptotic body per crypt section, following exposure to a low dose of ionising radiation. Generally, the highest frequency of apoptotic bodies will be observed around cell position four, where the stem cells are located and, in fact, we know that the apoptosis induced by a cytotoxic agent such as radiation is specifically associated with the actual, functioning stem cells, of which there are four to six per crypt. If we assume there are five stem cells per crypt and we cut a section through the crypt, we may have one or no stem cells on one side of the crypt section. If we use as the denominator all the epithelial cells in the tissue (crypt and villus), the apoptotic index would be about 1 percent. If we consider only the cells in the crypt, the index would become 4 percent. If we consider only the proliferative compartment of the crypt, the apoptotic index becomes 8 percent. However, the real apoptotic index for the apoptosis-susceptible cells (i.e., five actual steady-state stem cells) would in reality be 20 percent and could not be determined from a single crypt–villus axis, because this contains only one cell position four cell. The true index can be obtained only by counting a large number of longitudinal crypt sections and determining the

MEASURING THE LEVELS OF CELL DEATH (APOPTOSIS)

average apoptotic frequency for cell position four. Cell position four contains sixteen cells in circumference of the crypt, only five of which are probably apoptotic-susceptible cells. So, the best quantitative estimate one could obtain for the apoptotic index for the actual stem cells would be a value of 6.7 percent, whereas the true apoptotic index for the target susceptible cells, which cannot be specifically identified, would be 20 percent. So, in summary we have the problem that studies on tissue sections allow us to determine only a denominator based on the total number of cells or the number of actively cycling cells. What we really wish to define as the denominator is the size of the apoptotic-susceptible cell population.

The second problem associated with the interpretation of apoptotic index data is the fact that one has little reliable information about the duration of apoptosis, which determines in part the incidence figures. Clearly the number of apoptotic cells or bodies that one sees in a tissue section will depend on the time that it takes for a cell to go through the processes of apoptosis, fragmentation, phagocytosis, and absorption. In other words, the time that apoptotic fragments hang around in the tissue. This clearly varies from tissue to tissue and depends on many factors. In the small intestinal crypts some estimates have been made, based on limited experimental studies and these indicate that the duration of apoptosis is broadly speaking similar to the duration of DNA synthesis (i.e., the apoptotic process in its entirety takes between three and twelve hours). The complexity involved in determining the apoptotic duration is well-exemplified when one considers the small intestinal crypt. As previously stated, apoptotic events are commonly associated with the stem cell position in the crypt, at cell position four, immediately above the long-term residents the Paneth cells. Although this is true for radiation, the cell-positional distribution of apoptotic figures differs somewhat for other cytotoxic drugs.

In this tissue, the cells that remove debris (the litter) are the neighbouring epithelial cells. A few apoptotic fragments may be discarded directly into the lumen of the crypt and are lost via that process. Other fragments will be phagocytosed by the neighbouring Paneth cells, other stem cells, and neighbouring epithelial cells. Fragments phagocytosed by Paneth and stem cells will be digested by those host cells that themselves are not migratory, and so the loss of these fragments will be entirely determined by the rate of their digestion. Fragments incorporated into early transit cells will also be digested according to specific kinetics but they will also migrate at a rate determined by the migration rate of the transit population and will be removed by the two processes, digestion and movement. In other tissues, migratory macrophages enter the tissue and clear up the litter and so the time scale over which apoptotic fragments remain in the tissue will be determined by the speed with which macrophages enter, phagocytose the fragments, start the digestion process, and leave the tissue. These are all relatively understudied processes. However, some limited data are available from studies in the intestinal crypt. If we consider the time-course data from apoptotic induction (see figure 43), we can use the decaying portion of these curves to estimate the half-life of these apoptotic bodies, that is to say, the time that it takes for the burst of apoptosis seen at three to six hours to decay to half its value. This approach represents the only attempt to date to measure the duration of apoptosis in an *in situ* mammalian tissue. The results shown in figure 56 suggest that the half-life is of the order of three to twelve hours, when considering cell death induced by two different cytotoxic agents. This suggests that the duration of apoptosis is similar to the duration of DNA synthesis.

The next consideration in terms of interpretation of apoptotic indices concerns a statistical/technical problem that relates to the size of bodies being measured in sections of tissue. If we consider a tissue like the intestine and, in particular, the small intestinal crypts, proliferative and apoptotic indices are presented as numbers relative to the respective target population (i.e., the denominator in the index) and the denominator is usually determined by virtue of counting cell nuclei. The frequency with which one will see proliferative cells (as defined by mitosis) or apoptotic cells (as defined by apoptotic fragments) is determined by the relative size of the mitotic figure or the apoptotic fragment compared with an average cell nucleus (the denominator in any index). If, as is the case with mitosis, the structure is larger than an average nucleus, there is a tendency to overscore

MEASURING THE LEVELS OF CELL DEATH (APOPTOSIS)

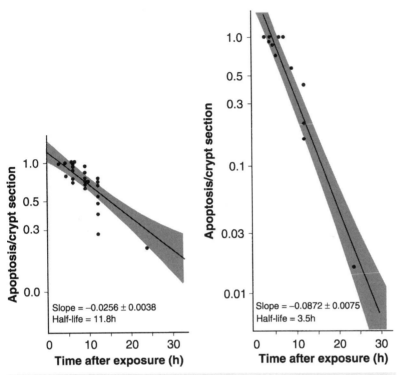

56. Decay curves for the removal of apoptotic bodies from the crypts of the small intestine. Pooled data from various radiation (left-hand panel) and drug (hydroxyurea: right-hand panel) experiments. Reproduced with permission from Potten, 1996.

the number of mitotic figures. If, however, the structures are smaller, as may be the case with apoptotic fragments, there is a tendency to underscore. The analogy that can be made here is as follows.

If one considers a table, on top of which there are three soccer balls and three table tennis balls, we have the same number of structures present, three per unit area of tissue, with the unit of tissue area being the tabletop. If we now consider cutting a section through the table with a chainsaw and then looking at the cut edge of the section for the entities on the table, there will be a greater probability that the chainsaw cuts through the soccer balls, than through the table tennis balls, so one will get a differential score for these two entities despite the fact that the real number of each is the same. This error in scoring can be corrected for quite simply by making some measurements of the relative sizes of the structures being analysed (the diameter of the soccer and table tennis balls). This size correction is rarely, if ever, applied to mitotic and/or apoptotic scoring. It has, however, been regularly applied to the scoring of surviving, regenerating crypts in the clonal regeneration assay for stem cell survival and functional competence in the intestine.

The next problem concerning apoptosis is the fact that one wants to obtain information about the number of cells that die in the tissue. The problem here is that these cells generally break up into a number of apoptotic fragments or bodies of varying but small size and it is these that are scored. In the intestine, we know that following a dose of radiation the dying cells can generate everything from a single large fragment to as many as twelve small fragments. This can be studied by looking at whole intact crypts and focussing through the crypts, looking for the clusters of apoptotic fragments. However, this is very labour-intensive and, hence, rarely undertaken. A distribution of fragment size is shown in figure 5. The problem here is really the reverse of the discussion concerning the size of bodies, because an apoptotic cell that fragments into twelve pieces may cover a space considerably larger than the average dimension of a neighbouring nucleus, whereas each fragment may be very considerably smaller. However, a single small fragment of such a cluster may be all that is seen, and hence recorded, for an apoptotic event. So scoring corrections that would theoretically be needed are much more complex and have never been adequately considered or discussed.

The final consideration in relation to scoring apoptotic bodies concerns the use of differing technical approaches and again generally speaking these have rarely, if ever, been validated one against the other and, indeed, it is difficult to decide on the gold standard with which all scoring should be compared. In the crypt, we believe that the gold standard is the morphological identification of apoptotic events, based on morphology in routine haemotoxylin-and-eosin-stained sections. These are easy to recognise and count, providing fixation, sectioning, staining, and microscope equipment are all optimum and that the observer has some experience in identifying dying cells, particularly in distinguishing certain apoptotic events from certain cuts through mitotic figures. Other techniques have been used, notably antibody staining techniques such as TUNEL. This is notoriously capricious and strongly dependent on the processing steps required. In the intestine, when this technique is optimised there always remain a number of false positives and false negatives. However, it is a useful and valuable approach for many other tissues. There are a variety of flow cytometric techniques that identify cells or cellular fragments with a sub-G1 DNA content. When these are combined with other markers they can provide reliable measurements of cell death in cell suspensions prepared from tissues, tumours, or cultures (see Chapter 2). However, some of the provisos outlined above still apply. There are a number of slightly less well defined approaches looking at cell death in cell cultures, which range from the simple loss of attachment (anoikis) and counting floating cells through to various approaches for detecting morphological changes in attached cells. There have been few if any attempts to validate the flow cytometric and cell culture measurements of cell death with *in vivo* morphological or staining techniques.

Further reading

Potten, C. S. What is an apoptotic index measuring? A commentary. *Br. J. Cancer* **74**: 1743–1748, 1996.

Potten, C. S., and Grant, H. The relationship between radiation-induced apoptosis and stem cells in the small and large intestines. *Br. J. Cancer* **78**: 993–1003, 1998.

Concluding remarks

Research into the process of apoptosis has expanded enormously after its first description in 1972 and a decade-long lag phase. Currently, a significant proportion of the cell biological literature is taken up with papers on apoptosis. Inevitably, there will always be some who like to quibble that apoptosis was discovered more than a century ago, but it is clear that much of the modern work was stimulated by the ground breaking paper of Kerr, Wyllie, and Currie in 1972.

There are still considerable problems associated with reliable and validated quantitation of levels of apoptosis and of how such data should be interpreted. The uncertainties here relate to the gold standard used for validation, the specificity of the assays and antibody techniques used for identifying the process, uncertainties concerning the duration of apoptosis in various situations and tissues, and the difficulty in defining the moment of onset and the moment of completion of apoptosis. In the tissues discussed in this book, the morphological changes associated with apoptosis still remain one of the most reliable ways for identifying the process, providing good fixation, tissue handling and staining techniques are used, together with high quality optics.

What is very clear is that apoptosis is an active, programmed process of cell deletion from tissues that plays a crucial role in maintaining the balance between cell production and cell loss that is an essential prerequisite of steady state in adult tissues of the body and that apoptosis also plays a crucial role during development. The homeostatic regulatory processes that determine the levels of proliferation and apoptosis in adult tissues ensure that manipulation or changes in one parameter invariably inversely affect the other.

Studies over the last twenty years have identified several of the crucial genes that regulate the apoptotic process and those that determine whether cells die or survive to proliferate. Early in these studies, *p53* was identified as playing a crucial role, and members of the *bcl-2* family soon became apparent as important regulatory genes determining cell life and death and the extent of the *bcl-2* family grew rapidly year by year. What remains to be understood is the intricate relationships between these gene families themselves and the many other gene networks and signalling pathways that are now beginning to be unravelled. Gene knockout experiments provided the rather surprising observation that in many cases there is redundancy and compensatory genes, which makes the unravelling of these processes difficult.

The susceptibility of cells of differing hierarchical status to have apoptosis induced, further complicates these studies. Stem cells are the most important cells, being the crucial originators of all cell proliferation in tissues, and stem cell biology has suddenly become a widely expanding area of cell biology. Ultimately, the tissue is totally dependent on its stem cells.

The small intestine is an excellent *in vivo* model to study not only apoptosis but the cell lineage dependence of this process. Stem cells can be identified in the small intestine by virtue of their position, and their susceptibility to apoptosis induction can be determined in a variety of situations. These cells are exquisitely sensitive to genotoxic damage and initiate apoptosis rapidly. This genotoxic damage is both *p53* dependent and *bcl-2* influenced. In this system, apoptosis also plays a role in the normal tissue homeostasis of stem cell numbers using *p53* independent processes. Apoptosis is also clearly important during embryological development in tissue restructuring – another situation where normal healthy cells are deleted. The DNA damage induced apoptosis in intestinal stem cells, which is p53 dependent, is part of a genetic protective mechanism that helps ensure that genetically defective cells are deleted, thus reducing cancer risk.

There is clearly much still to be learned about apoptosis, its regulation and biological significance, and whether we may be able to manipulate the process to our advantage in reducing cancer risk, increasing longevity, counteracting the effects of aging, or in designing novel therapies, particularly for diseases like cancer. It is hoped that some of the issues raised in this book will stimulate the next generation of research scientists to tackle some of these challenging issues.

Index

Numbers in bold indicate tables and figures.

β-catenin, 150, 167
Acridine orange, 55
Adenomatous polyposis coli (APC), 167
Agarose gel electrophoresis, 43
AIF (apoptosis-inducing factor), 102
Annexin V, 55
Anoikis, 58
Apaf-1, 102, 105
Apoptosis, 5, 17, **20–92**
   cytotoxic drug-induced, 162
   development, 185
   duration, 28–29
   extrinsic pathway, **109**
   hormone-dependent tissues, 185–187
   immune system, 187–188
   in embyronic development, 30
   in metamorphosis, 30–31, 184–185
   intrinsic pathway, **109**
   measurement of, 189–195
   morphology, 17–25, 36
   phagocytosis, 191–192
   radiation-induced, 155–160, 163–164
   recognition of, 42–57
   role of Bcl-2 in the large bowel, 164–167
   role of p53, 160–161
   role of, 23–25
   spontaneous, 152–156
Apoptosome, 105
Apoptotic bodies/fragments, 21, **22**, **28**, **57**, 61, 65, 190, 192–194
Apoptotic bodies/fragments, half-life, **193**
Apoptotic index, 190
Autoantibodies, 61, 64
Autoradiography, **80**
   principles of, **78**
   whole tissue, 78

Bcl-2 family proteins, 52–101, **108**
   Bax, 53, 55, 95, 99, 101, 102
   Bcl-2, 95, 98–99, 102, 154–155, 164–168, 172
   Bcl-2 knockout/null mice, 103, 165
   apoptosis in the intestine, **165**
   Bcl-w, 166
   Bcl-x_L, 99, 102
   Bcl-x_L knockout/null mice, 103
Bcl-2 family, Bcl-2 homology (BH) domains, 99–100
Bcl-2 family, mechanism of action, 102–103
Bromodeoxyuridine labelling, 79, **80**

INDEX

Caenorabdites elegans (C. elegans), 29, 62, 100, 104, 184
  ced-3, 100–104
  ced-4, 100–102
  ced-9, 100–102
  egl-1, 101
Cancer,
  Incidence in the intestine, 169
  tumour progression and role of APC, **168**
CARD proteins, 102, 105, 108
Caspase family, 49, 101–102, 108
Caspase family, sub-division, **101**
  Caspase-1 (ICE), 101, 103–105, 107
  Caspase-8, 105, 106
  Caspase-9, 102, 104, 107
  Caspase-10, 105, 106
Caspase substrates, **105**
Caspase-activated DNase (CAD), 104
CD31, 63
Cell cycle, 4, 73–75, **87**
  duration, 84
  G1 phase, 73
  G2 phase, 73
  M phase, 74, 77
  M phase, appearance, **74**, **75**
  quiescence, 84–86
  S phase, 73, 77, **82**
  S phase, identification of, 77–84
Cell death, analogies, 7–14
Cell division cycle 2 (cdc2), 88
Cell division, 71–75
Cell fate, 19, **32**
Cell lineages, **116**, **118**, **123**, 145
Cell Proliferation, 69, 75
Centriole, 72
Centromere, 72
c-fos, 95
c-jun, 95
Clonal/clonogenic regeneration, 142–144, 157
Clones, 6

Cloning, 4
Clonogenic/colony forming cells, 6, 126–129
Colony, 6
Colony-forming usints in spleen (CFU-S), 127
Cormack, James, 20
Currie, Alistair, 17, 33
Cyclin-dependent kinases (CDKs), 86, **87**
Cyclin-dependent kinases (CDKs), CDK2, 86
Cyclins, 73, 86–88
  cyclin A, 86
  cyclin B, 88
  cyclin D, 86
  cyclin E, 86
Cyclooxygenase (COX), 167–168
Cytochrome c, 53, 55, 102, 105

Death Domain, 106
Death Effector Domain, 106
Death Receptor Ligands,
  Fas Ligand (FasL), 106, **107**
  TRAIL, 106
  Tumour Necrosis Factor, 106
Death Receptors, 105–106
  DR5, 106
  Fas, 106, **107**
  Tumour Necrosis Factor Receptor (TNF-R), 105, 106
Death-inducing signalling complex (DISC), 106
Differentiation, 5, 16, 67–69
Dividing transit cells, 119
DNA damage, 31, 91–94
  cellular response to, **96**
  protective mechanisms, 170–177
DNA, 3–4, 21, 31
  degradation/fragmentation, 38–42, 155
  histones, 38, **40**
  laddering, 39, **41**

INDEX

nucleosides, 70
nucleosomes, 38, **39**
nucleotides, 70
Okazaki fragments, 70
purines, 70
pyrimidines, 70
repair, 83
replication, **80**, 170
strand segregation, 170–174

E-cadherin, 150
Endonucleases, 21, 38–40
   Endonuclease G, 102

Fas-associated death domain
   protein (FADD), 106
Fluorescein diacetate, 54
Fractional DNA content, 43, 88

Gadd45, 95
Granzyme B, 106–107
Growth, 2

Haematoxylin and Eosin, 56, **57**, **75**, 194
Heisenberg's uncertainty prinicple, 128
Hoechst 33342, 45, 55
Hypoxia, 8

Interleukin-6 (IL-6), 95
In situ end labelling (ISEL), 46–49
Ischaemia, 8

JC-1, 54

Karyolysis, 35
Karyorrhexis, 35
Kerr, John, 17, 33
Ki-67, 81

Leishmania amazonensis, 64
Li-Fraumeni syndrome, 91, 96–97

Macrophages, 62–65
Mdm-2, 92, 94, 95
Meiosis, 3, 72
Messenger RNA (mRNA), 68
MIB-1, 83
Micronuceli, 36–38
Mitochondria, 53
   outer membrane, 102
   VDAC, 102
Mitosis, 3, 71, 75, **77**
   anaphase, 73
   characteristics of, 75–76
   metaphase, 72
   prophase, 72
   telophase, 73
Mitotic cell death, 5, 36–37
Mitotic index, 76

Necrosis, 5, 18–20, 24–25, **32**, 37
Non-steroidal anti-inflammatory
   drugs (NSAIDs), 168

$p21^{WAF-1/cip1}$, 85, 86, 92, 95, 161
p53, 52, 85, 91–97, 106, 108, 160–161
   cellular role, **175**
   induced genes (PIGs), 95–96
   role in DNA strand segregation, 173–174
p53 knockout/null mice, 93
PCNA, 86
Percentage of labelled mitoses (PLM), 81
Percent-labelled mitosis, **82**
Phagocytosis, 22, 62–65, 191–192
Phagosome, 22
Phosphatidylserine, 55, 62–63
Pluripotency, 130
Poly-ADP ribose polymerase, 52
Programmed cell death, 5, 16, 32
Proliferating cell nuclear antigen (PCNA), 83
Propidium iodide, 45, 54
Pulsed-field gel electrophoresis, 43
Pycnotic, 35

INDEX

Retinoblastoma (Rb), 86, 95
Rhodamine 123, 54

Self-maintenance probability, 117, 126
Small intestine, 136–180
  apoptosis, radiation-dose response, **158**
  apoptosis versus crypt surviving fraction, **159**
  cell cycle time, 137
  cell lineages, **145**
  cell migration velocity, **141**
  cellular organisation, **138**
  columnar epithelial cells, 68, 136–139
  control of cell migration, **151**
  crypts, **47**, **57**, 65, 137, 139–151
  crypt surviving fraction, radiation dose-response, **143**
  crypt-villus axis, 140, 190
  cytotoxic drugs, cell positional targets, **163**
  distribution of apoptosis, **159**
  enteroendocrine cells, 68
  gene/protein expression, **178**
  gene/protein interaction, **179**
  goblet cells, 68
  Paneth cells, 68, 139
  stem cell distribution, **148**, **149**
  stem cell hierarchy, **145**
  villus, **47**, 137
Stathmokinetic process, 77

Stem cell plasticity, 132
Stem cells, 35, 115–125, 136, 139, 145–150, 154–155, 157–158, 160, 170–172, 190
  definition of, 120
  estimate of numbers, **144**
  control of cell division, **176**
  markers, 174–177
  repsonse to tissue injury, 124–125
Systemic lupus erythematosis, 64

T cell factor (TCF) proteins, 150
Terminal deoxynucelotie transferase-mediated dUTP-biotin nick end labelling (TUNEL), 46–49, 155, 195
Terminal differentiation, 69
Thymidine labelling, 77, **80**, **82**, 139, 157
TNF-R-associated death domain protein (TRADD), 106
Topoisomerase II, 40
Totipotentcy, 130–133
Tranforming growth factor-β1, 63, 64
Trypan blue, 54
Tumour necrosis factor-α, 63

Vital stains, 38, 54

WD-40 repeat, 102
Wyllie, Andrew, 17, 33